AF842831

Mobility and Environment

Corrado Poli

Mobility and Environment

Humanists versus Engineers in Urban Policy and Professional Education

A- 5'957'910

2012

Springer

2H 9348 2

Corrado Poli
www.corradopoli.net
policorrado@gmail.com

ISBN 978-94-007-1219-5 e-ISBN 978-94-007-1220-1
DOI 10.1007/978-94-007-1220-1
Springer Dordrecht Heidelberg London New York

Library of Congress Control Number: 2011929042

© Springer Science+Business Media B.V. 2011
No part of this work may be reproduced, stored in a retrieval system, or transmitted in any form or by any means, electronic, mechanical, photocopying, microfilming, recording or otherwise, without written permission from the Publisher, with the exception of any material supplied specifically for the purpose of being entered and executed on a computer system, for exclusive use by the purchaser of the work.

Cover image: © 2011 Jupiter Images Corporation Photo text: Aerial view of downtown Dallas, Texas

Printed on acid-free paper

Springer is part of Springer Science+Business Media (www.springer.com)

As always, there were plenty of practical reasons to justify the absurd and lead towards the impossible

Marguerite Yourcenar, Mémoires d'Hadrien

Foreword

> *"Ignoranti quem portum petat nullus ventus suus est"*[1]
>
> Lucius Anneus Seneca

> *"You don't need a weather man to know which way the wind blows"*
>
> Bob Dylan, Subterranean Homesick Blues

Goals and Structure

As we navigate the Internet, we will find hundreds, if not thousands of web sites proposing alternative traffic policies, new mobility patterns, and innovative transportation plans for European and American cities affected by congestion and pollution problems. Most of the proposals are very interesting. Some are brilliant. All refer to, or even invoke, the magic word: "sustainability". However, few of these proposals are actually applied. When, on occasion, they are put into operation, they represent a minimal part of the overall traffic and transportation policy which still concentrates on the construction of roads and heavy infrastructures. Why? In this essay I try to explain the gap between the widespread understanding of the need for change and the apparent difficulty of implementing radically new traffic policies.

About six years ago I wrote a pamphlet in Italian titled "*Rivoluzione Traffico. Meno mobilità più comunicazione*" that in English translates as "Traffic Revolution. Less mobility, more communication" (Poli 2006). I claimed that we need a dramatic change in urban mobility policies. My ambition was to propose an utterly new strategy. This new essay is the outcome of further research and proposes a thorough overturning of my previous approach. I no longer focus on alternative traffic policies designed to improve urban quality of life. I consider, instead, the problem of urban

[1] "He who knows not which port he is heading for, never finds a favorable wind".

traffic as a helpful starting point to understand the contemporary environmental and administrative crisis of cities. The revolution I call for is grounded on the idea that the difference between a superficial reform and a real revolution lies in the fact that the latter requires a different way of thinking which includes ethical and political considerations. Traffic policies are described as an advantageous battleground suitable to trigger social change. Thus, this book, contrary to the previous pamphlet, is not intended to deal with urban and traffic policies *per se*. Specifically it is an essay in environmental politics and ethics, in professional ethics, in education and in communication. The same considerations that I employ about traffic, may apply to most urban and environmental policy issues. This is why I introduce my essay elaborating on ethics and environmental policies, and more specifically with a critique of the sustainable development approach. Nonetheless, in the last sections of the book I have also listed a series of examples of new traffic policies and innovative tools to explain how a different thinking may also lead to innovative plans and practical innovations.

Since college, two authors, no matter how diverse from each other, have deeply influenced my way of thinking and my emotional/methodological approach to research: Gunnar Myrdal and Paul Feyerabend. The former synthesized the need of an ethical commitment in social research in his "*Value in Social Theory*" where he uttered that: "It might be useful to recall (…) that the social sciences have all received their impetus much more from the urge to improve society than from simple curiosity about its working. Social policy has been primary, social theory secondary. This holds true, of course, for the long ages from Aristotle onwards when the social sciences were merged into the general speculation which we have later come to call moral philosophy". The latter, in the introduction of "*Against Method*", his most famous essay on philosophy of science, declared that the "main reason for writing the book was humanitarian, not intellectual. I wanted to support people, not to 'advance knowledge". I wonder if I somehow contributed to the advancement of social science and knowledge, but with this book I honestly make an effort to support people and improve society.

Acknowledgements

I synthesize ideas drawn from professional consulting, political activity, and independent studies, as well as research and teaching I completed over thirty years in Europe, Australia and the US I also reconnect with the very intense seven year period I spent at Fondazione Lanza in Padua, Italy, during which, as Project Director, I promoted international studies and research on the political and ethical aspects of the environmental crisis.[1] The more recent origin of my research goes back to 2005, when I was asked by the University of Bergamo, Italy, to direct the Environmental Communication Laboratory at the Department of Communication, Language Science and Cultural Studies. Drafts of this essay have been used by students who graduate with Public Communication B.A. and M.A. degrees. A number of ideas reported in the book came from classes and seminars. I am grateful to my students for their active and enthusiastic participation. Part of the research and the writing was completed at the Johns Hopkins University Institute for Policy Studies where I spent the 2007 spring semester in the International Urban Fellows Program on a fellowship awarded by Compagnia di San Paolo, Torino, Italy. I returned to Johns Hopkins in 2008 as a Visiting Professor at the Department of German and Romance Languages, invited by Pier Massimo Forni. Having been part of the Johns Hopkins International Urban Fellows Program since 1979, I knew how valuable my time in Baltimore would be. Sandra Newman, at that time Director of the Institute, encouraged me to pursue my research goals and provided me with all the necessary academic support and precious advice.

[1] In that period I organized several seminars and three International conferences with the cooperation of the United Nations Human Dimension for Global Change Program, the University of Georgia at Athens (Ga.) and the University of Padova (Italy). Scholars such as the economist Kenneth Boulding, the Swiss theologian Franz Boekle, the environmental ethicists Holmes Rolston III and Dale Jamieson, the epistemologist Kristin Shrader-Frechette and the Chinese economist Yu-Shi Mao took part in the program. See Poli (1994a, 1994d), Poli and Timmerman (1992, 1993), and Ferré and Hartel (1994). I have been educated in approaching problems comprehensively since the very beginning of my studies in politics and geography. This is why Fondazione Lanza hired me for the position. Thus, I had the opportunity to coordinate an interdisciplinary group of highly qualified scholars, who were aware of the epistemic issues concerning their studies and skills. Hence, I was stimulated to understand diverse disciplinary languages which were correctly and properly spoken.

Contents

Chapter 1
Introduction: Technicians and Humanists in the Environmentalist Debate on Mobility and the City

Hence, I concluded that trouble is
inevitable and the task, how best to make it,
what best way to be in it

Judith Butler (1990)

Abstract In some situations we wrongly use technology and engineering in order to comprehend and solve crucial contemporary social and administrative problems. This is especially true regarding urban traffic and its environmental consequences. For many decades, civil and transportation engineers were the only professionals substantially in charge of traffic policies. Because of the apparent failures in coping with urban traffic, it is appropriate to analyze if other disciplines and approaches may be more successful than the monopolistic civil engineering approach. The environmental question, which is a priority in European and North American urban policies, includes relevant epistemic issues that need to be considered in decision-making. Hence, it is necessary: (a) to define how professions and professional guilds affect in reality the everyday setting of the traffic problem; (b) to discuss how urban traffic policies are consistent with different basic interpretations of the environmental crisis; (c) to argue about the structure of the contemporary urban settlement and see whether mobility is a solution to the current problems, or a pathology to be cured.

Keywords Engineering vs. humanities in decision-making • Professional ethics • Education and training-curricula • Environmentalism • Transportation vs. communications • Social change • Cultural revolution • Urban democracy • Citizen participation • Epistemology

C. Poli, *Mobility and Environment: Humanists versus Engineers in Urban Policy and Professional Education*, DOI 10.1007/978-94-007-1220-1_1,
© Springer Science+Business Media B.V. 2011

Technicians and Humanists

My brother claims that nobody has ever seen me holding a screwdriver. This is not true. Occasionally, I do dabble in minor carpentry and electrical work around the house. Nonetheless, it is a fact that house maintenance is neither my favorite activity, nor the one in which I have proven talent. My embarrassing lack of skill with screwdrivers and hammers is a source of frustration for me, because I am truly fascinated by technology and engineering. I really admire people who possess the skill and manual ability to fix things; those who can improve their lives with simple, but efficient practical solutions. At a more sophisticated level, I admire dexterous engineers who can invent techniques and mechanisms that solve problems and simplify complex situations. New techniques can often resolve many problems. They have throughout history and will continue to do so in the future. In fact, the whole purpose of engineering is problem solving. The time that engineers save not questioning the deeper meaning of life, the uncertainty of reality and political issues, can be profitably applied to tackle specific problems. Thus, they receive great satisfaction and maintain a psychological balance that philosophers and humanists rarely enjoy.

However, in some situations we wrongly use technology and engineering in order to comprehend and solve crucial social and administrative problems. This especially includes urban traffic and its environmental consequences. Here my criticisms are not directed at engineers and technicians. The real target is the political decision-makers whose responsibility is to assign engineers and technicians well defined tasks. People who hold political offices need to properly manage the professional services employed in planning and decision-making processes. I suggest that decision-makers, social scientists and humanists, should play a leading role in mobility and transportation policies, a field currently monopolized by technicians.

Technicians and engineers are trained to solve problems, not to identify and set them. At best, they may reformulate problems in new ways to facilitate innovative solutions. Epistemologists and scholars in professional ethics understand this fundamental separation of functions, as do engineers who happen to be also educated in the humanities. There are enlightened engineers and politicians, open to innovative solutions that go beyond technical challenges and lead to broad political action. Anyone can be enlightened and open-minded no matter what profession they practice. However, in general, engineers' education does not systematically include political and social studies. Thus, it is necessary to distinguish between those who choose and set the problems and those who are responsible for practically solving them. This is essential for the progress and development of appropriate techniques. As the environmental crisis worsens, I have become increasingly perplexed by the fact that civil and transportation engineers are the only ones substantially in charge of traffic policies.

As we can expect, there is an abundant literature about engineering and epistemology (see, e.g., Royal Academy of Engineering 2008). Many scholars have also analyzed how decisions about transportation and infrastructure are made in real life. In *Transport of Delight. The Mythical Conception of Rail Transit in Los Angeles*,

Jonathan Richmond (2005) presents a detailed description of the decision-making process and the confused intertwining of political and technical considerations. The author explains how things really happen and locates the apparently dominating technical discourse in the real political process. Richmond-like analyses are not pondered, as much as they should be, by decision-makers nor are they studied enough in the regular training curricula for planners. Even less are they bookstore best-sellers. It is a shame because this kind of analyses may unveil the inefficiency of the current urban policy regarding traffic and transportation. But these books demand the readers to think and re-elaborate, which is a difficult and tiring activity. It is much easier to absorb a simplified explanation based on shared paradigms, biases and dogmas. Because the general approach to traffic problems is not questioned from epistemic and social points of view, therefore no reforms in education, professional training and policy-making have yet been effectively proposed and discussed.

My essay concerns urban policy and planning. Some parts of it ought to be read as an attempt to re-evaluate the role of engineers and technicians. These professionals may find new significance by accepting the challenge to face new problems and/or the old ones, reformulated. In other words, we should no longer have civil engineers to be merely basic craftsmen with limited skills. Instead, they should be viewed as ingenious innovators. Professionals in engineering should be relieved from the burdensome considerations of the social and political consequences of their job. They should refrain from this because others, more qualified, are in charge. If we employ engineers to correct flaws in policy implementation – or worse yet, allow them to abuse the political scene – they lose our professional and ethical consideration. Mainly, they become ineffective. Being ineffective is the worst insult to an engineer.

Hence, my essay is a call for more thinking about decision-making processes so as to advance some suggestions on how to break the impasse. The problem of the skills we need to innovatively approach an urban and transportation policy based on environmental concerns is clearly proved in the shallow review of Richmond's essay by the distinguished economist Peter Nijkamp in "Planning Theory" (2006). Nijkamp fails to understand that Richmond's approach is based on a completely different communication language and in a quite assertive manner defines the book "wordy" and lacking structure.[1] Miscommunication is worsened by the unjustifiable arrogance of some scholars over others who try to adopt a language out of the mainstream. Many scholars do not know appropriately what social scientists, ethicists and philosophers' competence is and also avoid arguing about what engineers, technicians and scientists can do and what they cannot (and mainly shouldn't) do. In this respect I still find Gunnar Myrdal's 40 year old critique of the western standard social and economic theory applied to "backward" countries in order to start

[1] I thank Jonathan Richmond for having discussed with me Nijkamp's review. I need to add that most of the reviews of Richmond's essay have been positive and *Transport of Delight* has been appreciated in diverse cultural milieu.

development very inspiring. Myrdal (1968, p. 6) calls for more attention to the biases of experts and, not without irony, writes that …

> … we economists and other social scientists are now studying intensively how people behave, and how they are motivated and then conditioned both by their inherited constitution and by their environment. We are interested in the selective process operating on the young as they find their way in life and are guided into different occupations. We are examining the formation of opinions and attitudes, especially decision-making by public administrators, business managers, employers and employees in the labor market, and political leaders and their followers. We are observing how people spend their leisure time, how they marry and pursue family life how some become criminals, vagabonds, or prostitutes. In short, we are concerning ourselves with human behavior and motivations, in whatever profession, social class, or geographical location.
>
> Only about the peculiar behavior of our own profession do we choose to remain naïve. How we as scientists operate in seeking to establish knowledge is largely shielded from the searchlight of social study. But, surely, though we are seeking truth, we are not less conditioned by our mental make-up and the society in which we live and work than are other men. … Our research interests, the particular approach we choose, the course we follow in drawing inferences and organizing findings, are not determined by facts and logic alone. … And yet, although literature and art have long been considered in relation to the psychology and the environment of their creators, our writings have not been.
>
> Our lack of curiosity about our own peculiar behavior as researchers should be surprising. As a group we are certainly as interesting and important to the dynamics of the social system as are maladjusted girls, new immigrants, and other special groups in society that we are studying more and more intensively; we perhaps rank with business managers, professional politicians, or creative artists. Our behavior can be easily ascertained from our writings.

As a matter of fact, if we do not inquire about ourselves, our biases, preferences, hidden interests, we will never be able to carry out a meaningful critique of the structure of the problems we want to face: wherever we look, we will see anything but ourselves.

An Environmentalist Debate

I intend to participate in the environmentalist debate and aim at overcoming obsolete and harmful concepts such as sustainable development (see Chap. 2). Until 20 years ago, and even more recently, the very word "environment" evoked a progressive and even revolutionary vision. At that time some political leaders, including conservative scholars, dared to deny the very existence of an environmental problem. If we consider all the consequences of the contemporary mobility patterns of people and merchandise, we find that transportation and mobility at global scale are the most polluting human activities and produce the most detrimental impact on earth. If we limit our attention to urban mobility, the environmental impact produced by frantic, senseless mobility becomes the major issue in urban environmental policy. These environmental issues not only include air pollution from gas exhaust, but also the need to build extensive highways, railroads, parking garages, and to produce, and subsequently to dispose of, an incredible amount of transportation components. This is why transportation and mobility policy is such a significant part of overall environmental policy. Yet it is not the only reason: contemporary insane mobility patterns are the consequence of an inefficient organization of urban life (see Chaps. 5 and 6). Rather than being a disease itself, traffic should be more appropriately considered as a symptom of seriously ill cities. Because the contemporary world is thoroughly urbanized, stemming from the mobility crisis we can argue more in general about contemporary society.

Mobility and the City

We define mobility as the physical movements of people in a region. Transportation regards the means to make physical movements possible. Traffic is the specific mobility in an area and refers to the flows of people and goods on transportation infrastructures. Traffic congestion is the pathology of mobility which is now a quasi permanent condition, a toll we have to pay to receive some unproven benefit (see Chap. 3). The centrality of mobility and transportation in urban policy and planning is attested by a more updated definition of the city. For almost the entire course of human history, cities have been "places where *some* people lived." In the twentieth century they became "places where *most* people lived."[2] Finally, in the last 50 years, cities are better defined as "urban areas where people move around." In current conversation, for every action we are going to do, most times we add the verb "to go." Movement has become an essential part of many actions we perform. So we do not work, but *go* to work; we do not study, but *go* to school. We even *go* to exercise … driving sometimes many miles to *go* to the sport facilities … in order to run more miles. Cities have morphed into a thoroughly urbanized milieu, as it has often been repeated. Thus, discussing urban and cities is the same as speaking about society in its entirety.

Architects and urban planners design cities focusing on buildings according to millennia-old approach. They take into consideration roads, streets, and parking, but it happens after and not before the design and the building-centered functional plan. If cities are primarily places "where we move around," the approach should be the opposite. Following an Imre Lakatos' concept (Currie and Worrall 1980),

[2]Kingsley Davies produced the fundamental demographic studies about urbanization process in the 1960s (1969, 1972). Now, there are still researchers who update Davis' data using a similar methodology which in the meanwhile has become meaningless. While in the past it was possible to trace city borders and thus calculate the number of people living in urban centers ranked by size, now this is not possible anymore. In fact, urban settlements have spread in larger and larger areas, but data still refer to administrative units whose borders were typically established in a pre-urban situation. Moreover, contemporary urban lifestyle is extended to most of the globe and all the people living in western countries adopt an urban lifestyle in terms of consume and production, no matter if they live in a low-density area, or even in the woods: they proudly drive a four wheel drive pick-up to shop global merchandise at the closest shopping mall's grocery store.

i.e. "rationality works much slower than most people think and, even then, fallibly", we must accept that there is a lag between the appreciation of a technique and its widespread actual application. Most of the time innovative industries might quickly reconvert their production and apply new technologies, but people's entrenched attitudes, and the production system's aversion to change, restrain the conversion process. Sometimes, everything seems ready for new ideas and technologies to be applied; there's a lot of speaking and planning about them, but a real implementation of them never actually takes off. In psychoanalytical terms we may suggest that people consciously adopt change, but their unconscious attitudes still determine their innate behaviors.

I think we should go back to cities intended as "places we live in" and we should plan to reduce movements both within and between cities. Thus, we cannot separate traffic and transportation planning from city policy as a whole. As a matter of fact, this is anything but new, having been proposed by Jane Jacobs and her numerous followers as long ago as the fifties and even before by Lewis Mumford and many geographers and planners. This idea has been constantly repeated though never (or rarely) implemented. This book focuses on the political consequences of the interpretation of the city as "a place we live in". I argue that the application of telecommunication and information technologies will have essential effects on physical mobility and hence on traffic patterns. They will also hasten the transformation of urban society and politics, not to speak of a possible major shift from the construction industry to information technology industry and communication processes. The widespread diffusion of the internet is only a ten year old phenomenon, a time too short to conclude whether it has already changed people's mentality and production structure. Although we spend hours working and amusing ourselves on line, and cannot survive without a cell phone, we still live and think as if we need more physical mobility than is logically necessary. This applies indifferently to both old and new generations. It is not a matter of mentality and attitudes that can presumably vary according to age groups; it is a consequence of the old mobility oriented urban organization that still affects everyone. In fact, our mental patterns and social organization have not yet been subverted by the all-encompassing communication technology. Something similar had happened when mass-motorization boomed: it was easy to buy a car for everyone, but it took much longer to have society, infrastructures, and people's attitudes adapt to the new lifestyle made possible by easy mass mobility. In the mid-to-long run, new technologies will definitely affect people's behaviors, industry and settlement patterns. Small towns and local communities that carve their space inside never-ending metropolitan areas, will become more competitive compared to big cities in terms of quality of life and wealth production. Big cities will not be able to solve traffic problems – as well as many other typically metropolitan crises – created by the obsolete social organization established by twentieth century technology. Therefore, progressive change will have a chance to happen in small towns, in decentralized areas, or in small communities. They enjoy the advantage of participating both in the global system and also having the possibility to stay separate from it thus being free to develop an alternative lifestyle and alternative politics. Transportation and mobility policies have a crucial role in this development since large metropolises are characterized by generalized mobility. On the contrary, in small towns and in communities, mobility patterns rely on stability and on selected and finalized movements. Moreover, urban transportation is expensive in large urban areas. It requires investments that in smaller compounds may be used more effectively for other purposes rather than moving people around.

Small towns' and communities' social and political milieu is a promising battleground for fostering change. Notwithstanding this strategic advantage, we still need to overcome the inertia of a conformist thinking in order to enhance more effective change policies. To transform the ongoing technological revolution into a political change process, I criticize the current political system which supports the mobility and transportation business (Chap. 4); then I will propose actions related to professional and science policy (Chaps. 5–10).

The Post-Global Condition

All this happens when the globalization process is approaching an end, when a global system has ultimately taken over. Somebody has already spoken of a post-global condition. In the last 40 years every definition of the current times could not help being preceded by the prefix "post", namely a reference to the past: post-modern, post-industrial, post-structuralism, deconstructionism, post-democratic, post-political, and more. "Post-global" is just one additional entry in the list. For four decades we have been imagining the future by turning towards the past. It is the proof of a decadence which includes the aspiration to something completely new: "Minerva's owl only flies at dusk"! But dusk is dusk and night isn't hurrying to fall: hence "the old is dying and the new cannot yet be born … in the interim, a great variety of morbid symptoms appear" (see inscription, Chap. 2).

Two strategies are currently available. We may either try to advance on the usual familiar paths, or we may focus on the destruction of the past.[3] It is time to come forward with a third option, namely propose a "pre-something". By arguing about traffic, I do not focus on the post-global city. Even less do I propose any good-for-all new solution to current urban traffic problems. Every practical solution is conservative *per se*. "Solutions" are thought to fix the system's flaws for good, hence to

[3] We do not need to go in depth into a sound philosophical argument that others can develop better and devote themselves to. However, let me say that this argument echoes the classic debate between Habermas' Enlightenment as an "unfinished project", and Adorno-Horkheimer's "dialectic of Enlightenment". The second position implies that Enlightenment *per se* will bring us to totalitarianism; the first implies that we still have to proceed on the usual path. You may call my statement a pop version of this noble debate.

preserve it. Solutions are motionless, interpretations are dynamic as they raise new questions. I will propose an interpretation and a direction for research and political action. With this essay I do not elaborate completely original ideas, but I contribute by planting seeds for a radically new approach to come.

A story and the following paragraphs will help to clarify what I mean to convey in Chap. 2 and throughout my essay. The forest having caught fire, all the major animals are helping to extinguish the flames. Among them, a humming-bird is bringing water though proportioned to her minuscule size. An elephant, who is effectively spraying water over the burning grass with his huge trunk, asks her:

> "What are you doing?"
> "I'm extinguishing the fire!", the humming-bird shouts convincingly.
> "You are only bringing a few drops of water!" the elephant says.
> "Doesn't matter, I'm doing my part", replies the humming-bird.

This story is often told to teach children that we all should contribute as best as we can to the general well-being. The moral is acceptable and a good one, in principle. Nonetheless, we may also head to a different conclusion: the story approves of people who act without thinking. The allegedly laudable humming-bird refuses to act politically and prefers to appease her conscience by doing something practical and useless, but mainly safe for the conservation of order. Eventually, the big animals, who hold the power in the forest, prefer to have humming-birds busy doing something useless but harmless rather than inappropriately questioning whose fault it was that the forest caught fire. Or, even more dangerously, challenging somebody else's power. This is what often happens when environmentalists are allowed to develop irrelevant projects – or even allowed to protest disjointedly – instead of fighting against the system.

The classic nineteenth and twentieth century revolutionary approach – first seize the ruling class power and then proceed to social transformation – is no longer popular among radical thinkers and activists. The most popular contemporary revolutionary movements – such as the Zapatistas, the "Movement of Movements", "We are Everywhere", "Other Economies Are Possible", "One No, Many Yeses", "Life after Capitalism", the "Reclaim the Streets" movement and so on – all share the gloomy belief that they will never be able to overthrow the capitalist system and the dominating neo-liberal order. As a consequence, they adopt what Bennett defines as the Zapatistas' "playfulness and humor which would toss us onto the terrain of the possible" (quoted in Gibson-Graham 2006, p. 196). This position is doubtlessly realistic, too bad that it is conservative instead of revolutionary as it would like to be (Žižek 2006, p. 237). All these revolutionaries (who'll never dare to start a revolution and consider themselves defeated from scratch) crave for, is to carve a niche where they can hide and live as they like. They claim to be a virus that in the long term will infect the global system and drive it to its own destruction. Waiting for the long term to happen, in the meanwhile they can have a lot of fun safely enjoying their innocuous revolutionary identity. Unfortunately, we have gone over this phase without making any real step forward in the direction of spreading new contaminating ideas. On the contrary, the movements have been marginalized and are brought into the limelight occasionally, possibly when they are instrumental to the neo-capitalist order. If we think that revolutionary change is a serious ethical issue, we need to identify new political strategies capable of involving at least part of mainstream public opinion. It is useless to preach to the choir as we have been doing for years; rather, we should move to Ninive to convert the unfaithful. If the movements really want to contaminate society's beliefs with new ideas, they must create a contact with the system they want to overthrow. Otherwise they are neither political nor ethical and end up drawing together merry bunches of good old fellows who enjoy life as much as they can! It is useless implementing minor projects instead of dealing with your neighbors, no matter how differently they think. Political – hence relational – action is the main and only possible source of change. Who fears change, curbs politics. The only hope of a better quality of life is in a renovated political action that must be based on a now-missing affirmative political thought.[4]

The Urban Secession

To properly face the mobility problems, I propose a mobility revolution based on (a) an urban secession and (b) on a reshuffling of the professional competences of those who are customarily put in charge of solving them (Chaps. 9 and *passim*). The word "secession" should not be taken literally. Secession is meant as: (a) a way of thinking and describing urban geography; (b) a leading idea for urban plans and policies; (c) the corner stone for revitalizing urban democracy; (d) the starting point for new administrative and fiscal partitioning; (e) eventually, a new form of self-government of federated cities/communities.

My secession and mobility revolution are symbolic. I claim that we need to go back to the small scale if we want to solve traffic and urban problems – the opposite of what has been done in the last centuries. This does not mean that we should imprison ourselves in culturally closed communities. In fact, the possibility of moving from one city to another – and the very existence of "another" city – is exactly the opposite of the uniqueness of the global condition. The so-called provincialism is a conception of the world that does not consider anything but one's own community. Thus, there is nothing more provincial than globalism. Out of the individuality of the globe, there is nothing else. I claim for a political and technological approach – technology is itself political – which: (a) allows the development and the implementation of new ideas; (b) creates a competition between different life styles and technical solutions; (c) reconstructs a relationship between land and people that we have lost in the global world. We should conceive progress

[4]This is a basic conviction that has shaped most of this essay. Admittedly it is not completely and explicitly elaborated, but the foundations of it are exposed in Chap. 2. I will return to it in a forthcoming essay.

as an improvement in quality of life obtained by utilizing local resources to a larger and larger extent. Among these resources, the pivotal one is a democracy that can be effective when there is a direct relation between individuals and between people and nature.

A Call for a "Cultural Revolution"

In the following chapter, I argue for the opportunity to position the relation between humans and nature at the center of the current political debate. Such an approach would affect traffic policies dramatically since it would alter the evaluation of the mobility needed. Being aware of the paramount complexity of positioning our political focus on the relations between humans and nature, I give my contribution by proposing some ideas on how to subvert politics and decision-making related to urban traffic policies. I indicate a possible discussion arena and prepare the ground for more elaborate thought.

The problems of traffic congestion, air pollution, and the low quality of city life will not be solved if we do not radically change our way of thinking and acting. This does not mean that we should change the current situation overnight, no matter how dramatic the need. Even less would I claim we need to enforce change with authoritative methods. However, it is time to think differently, identify new processes of change, and, eventually, propose some practical solutions. The practical solutions I propose are intended as tools designed primarily to promote social and political change. They are not the ultimate goals neither of my study nor of urban policy. In Chaps. 3 and 4, I describe some entrenched prejudices and ways of thinking. They have been reviewed systematically by many authors to whom I refer in the bibliography. Hence, in these chapters I give an overview aiming at raising new questions and providing a new perception of traffic and urban problems. I focus on some ethical, political and professional aspects aimed at introducing Chap. 11 in which I advance a planning procedure and go on to describe some possible projects (Chaps. 12 and 13).

We should relinquish the century-old expectation that science and technology can provide decisive answers. No one can presently claim enough authority to affirm their truth, in whatever situation: be it political debates, neighborhood meetings, or public hearings. Therefore, citizens and policy-makers do not create an inconsistent discourse only because they refuse to communicate and interact, but since they speak different languages and do not understand each other. The consequence of the approach I am proposing in this essay is that, in order to make a decision, we should emphasize the relations among stakeholders, who each bear contrasting interests and – what is even more inhibitory – different values. Current technique-centered decision-making is inefficient since it bears a hidden violence based on the assumption that there is only one possible best solution that matches anybody's needs, everyone's preferences. Instead, we should accept that rationality requires more emphasis on the process, rather than on the solution. This approach would *per se* be rational, albeit it deals with apparently irrational subjects and behaviors. If we agree to this approach, then social, communication, and political science skills should play a major role in planning, starting at the very beginning of the decision-making process.

The Centrality of Communication

In the customary decision-making process, broad consensus and fair decision-making are beyond the core of environmental policy making, although they are regularly (and desperately) called upon. On the other hand, consensus-building and decision-making are more important than any supposed rational, technologically advanced, good-for-all solution which is legitimized by science or scientific organizations. We can assume that a rational approach exists only if there is enough rationality in the communication and decision-making process. In this respect I refer to Shrader-Frechette (1991) who proposes a "procedural rationalism" approach in risk and decision-making assessment. Shrader-Frechette significantly takes into consideration what she defines as "ethical weighting techniques" that she considers necessary to introduce methodological reforms in decision-making evaluation. A similar position has been recently reported by Sheila Jasanoff (2010) who notices how a significant shift has occurred in the last 30 years: while in the past decision-makers and public opinion relied almost completely on expert judgment, now there is "recognition that public consultation improves the quality and the acceptability of expert judgment".[5] This case can be applied to the traffic plans designed by civil engineers who present their solutions as mere (and only scientifically disputable) technical documents.

Communication and participation, however, are anything but simple notions. Some basic background in political philosophy and in political studies is necessary to understand what communication and participation are exactly and what are the methods and the consequences of policies applying participation and communication tools. Communication methods are intertwined with citizens' participation in the decision-making process from both a practical and theoretical point of view. In practice, communication is more political than a supposedly neutral and objective expert evaluation and decision. Nonetheless, there are helpful technical methods – sometimes defined as social and institutional engineering – designed to ease communication among groups of people bearing diverse values, goals, and

[5]"A 1983 and a 1996 report of the National Research Council bookended the turn from integrity to accountability. The first (National Research Council 1983), recommended that the largely scientific exercise of risk assessment should be separated as far as possible from the political and value-laden task of risk management. The chief purpose was to protect science against possible biases. The second (Stern and Fineberg 1996) concluded that risk analysis should be seen as an intertwined analytic deliberative process, requiring repeated public consultation even in the production and assessment of scientific knowledge. Here, there was recognition that public consultation improves the quality and acceptability of expert judgments" (Jasanoff 2010, see also Zimmerman 2010).

expectations. If we are clear about the political content and consequences of the communication process, we can more easily adopt reasonably neutral, fair, and objective communication techniques (see Chaps. 10 and 11).

All urban traffic projects and plan implementations considered communication and participation at some point along the way (Chap. 7). In this essay, I claim that a systematic and seriously designed communication and participation method should be included from the very start of the planning process, and should be considered an essential part of the plans. Many of us have always regarded participation as an ethical imperative of democracy. Today, however, it has also become a necessity for rapid and effective decision-making.

Participation has many different meanings and applications (Nagel 1987). In general, when calling for more participation in the planning processes, the intention is for a more direct involvement of the citizens in the design and decision-making process. Although participation can be facilitated by neutral group management tools, it is essentially a political matter and should be treated as such. We will never have real citizen participation, i.e. democracy at large, if the citizen's involvement process is not strictly connected with official deliberations in governmental institutions.

Procedural vs. Scientific Truth

Sheila Jasanoff in her *Science at the Bar* (1995) also reports on an interesting case.

> A 1993 decision by a New York State's highest court held that a claimant could seek damages for a drop in property values caused by public fear of a right-of-way for a high voltage power line. The claimant was not required to prove that there were medically or scientifically reasonable grounds for the phobia concerning the effects of electromagnetic fields. The court felt that the economic question of the loss in market value could be resolved without being 'magnified and escalated by a whole new battery of electromagnetic power engineers, scientists or medical experts.

This episode – and some others also described by Jasanoff – highlights the growing importance of communication in decision-making. The authority of traditional scientific institutions, which in the past guaranteed veracity – or at least legitimacy – has been undermined. The possible legitimate coexistence of different paradigms has of late been addressed in a number of philosophical and epistemic studies. However, a well-defined rationality that applies to, and is accepted by everyone, is still the dream of many "self-defined" practical men. They are best described in J. M. Keynes' celebrated quote as believing themselves "to be quite exempt from any intellectual influences," but being actually "the slaves of some defunct economist." Rational argumentation and scientific language are still used in most public hearings by both citizen activists and transportation administrators, even when they advance two opposing and incompatible truths.

The increasing importance in contemporary society of communication *vs.* objective truth was observed also by Daniel Goleman (1995), a controversial psychologist who studied what is called "emotional intelligence." In Goleman's opinion, empirical evidence demonstrates that employees with a higher emotional intelligence are more successful than those with a higher IQ. Evidence and counterevidence has been brought by both the IQ supporters and by Goleman. What really matters here is that companies are now keen to hire good communicators and people who can easily relate to each other, rather than the most brilliant "objective" problem-solvers as in the past. If this is true in the market competition, we should consider the possibility to approach the traffic problem from a social perspective rather than with physical constructions that are the unavoidable outcome of engineers' rational plans.

Conclusion

All these points, which will be developed further in the following chapters, stand at the foundation of my education and of my professional and political practice. I believe that they should become a political agenda and help promote the democratic principles of our society. Democratic decisions are the "right" decisions and are to be considered also the most effective for everyone. Nonetheless, people who gain some power in a community (or in a country) may declare that they endorse these same citizens' goals and this is why they do not have any real interest in sharing power with the communities. This is not merely a matter of malevolence, corruption, or lust for power. Something else drives people-in-power in many circumstances. In fact, we must recognize that it is clearly inconvenient for average politicians to favor a process that will deprive them of a share of their power, and so it is natural for them to try to hold back citizen participation. Often politicians are driven by the honest conviction that they are acting for the common good. Most people cannot help being average or close to average, otherwise the very notion of average would be false. Most people are anything but mediocre. A noble politician, a real statesman, inspired only by ideals of equality and democracy, would gladly relinquish some power in obedience to those ideals. But we cannot ask donkeys to gallop. Thus, if we want to make democratic, ethical and effective decisions, we need to devise the most appropriate strategy. It means that we need to persuade average politicians to regard the participation process favorably and to create compelling reasons to allow citizens' involvement in the decision-making process. I admire enlightened and noble statesmen and politicians, but I trust the mediocre more, provided they operate in democratic and well functioning institutions. One of these reasons – probably the most appealing also for the most conservative politicians – is certainly that a decision supported by a large consensus can be more easily and quickly approved. We hand over power in exchange for a precious resource in politics and business – namely time. These considerations apply to our traffic issue, maybe more so than to other policy problems. As long as leaders are involved in government, time is a precious resource that a well-structured decision-making system can help to create. In the first steps of our "mobility revolution" we can introduce innovative approaches by negotiating rapid decision-making with consensus on projects that will prove strategic in priming a real change in the long term.

Chapter 2
Sustainable Development: From Fallacy to Fraud*

It is by this that Rubens proves himself great
and shows to the world that he, with a free spirit,
stands above Nature, and treats her conformably to his high purposes.
... But if it is contrary to Nature,
I still say it is higher than Nature ...

Wolfgang Goethe (reported by J. Eckermann)

Faust is dead!

Guenther Anders

The atomic bomb has made Goethe unlikely

Karl Jaspers

Abstract In this chapter I frame urban traffic policies within a general approach to environmental policy. As early as 1992, scholars and western public opinion were taking steps towards constructing an environmental ethic fitting for an active green political platform. The 1987 U.N. Sustainable Development compromise has inhibited this process and transformed the environmental problem from a political and ethical topic into a technical, economic and scientific one. I define four levels of environmental consciousness that may inspire urban traffic policies. Then, I propose a strategy to break the impasse generated by the Sustainable Development pseudo-ideology. My assumption is that we need to find another dialectical alternative to the

*To a large extent this chapter reports on an essay that was published in *Human Geography: A New Radical Journal* (Poli 2010). I thank the editor Richard Peet for having allowed me to republish part of it in this book.

C. Poli, *Mobility and Environment: Humanists versus Engineers in Urban Policy and Professional Education*, DOI 10.1007/978-94-007-1220-1_2,
© Springer Science+Business Media B.V. 2011

neoliberal order which has substituted for the defeated Communist option. The U.N. Sustainable Development compromise – which was signed before China and India's economic boom – is no longer suitable to stand as an alternative. At the same time, other ideologies are becoming possible alternatives (threats) to Western society. A radical change in urban and traffic policy can be the starting point of a challenge to both neoliberal capitalism and other possible unwelcome alternatives. I propose an environmentalist option based on responsibility and citizenship which would substitute for the traditional two century-old political debate hinged on freedom and justice.

Keywords Environmental policy • Environmental ethics • Philosophy of science • 'Green' movements

Introduction

Today, the endangerment of nature and the growth of environmental risks are generally acknowledged. Everyone is occupied with the environment, but only a few put the relation between human beings and Nature at the center of the political debate, and call for substantial political and social change to confront environmental problems. I develop this argument in this chapter. It is not specifically related to traffic policies and is meant to define the political and scholarly framework in which I roam.

Sustain or Progress?

Would you be happy if, being young and in love, filled with enthusiasm and expectations, your beloved partner replied to your proposal to pursue a lifelong relationship by saying: "It's ok, I think we can have a 'sustainable' relationship and our ultimate goal will be to make it last as long as possible, no matter how we feel and what we do. Hence, don't ask me to change any of my routine and I'm not going to do anything to deal with my possible shortcomings." You would probably not appreciate such a response, unless you were so dejected and your life was so miserable that you couldn't even conceive any real improvement in your gloomy existence. Surely you would prefer a response along the lines of: "Yes, I am going to share my life with you and this relationship will help us both to realize a real improvement in our lives. Together we might even be better off, but what really matters is our emotional fulfillment. Our lifelong relationship will make us better human beings and we will fulfill our personalities and satisfy our everyday needs. We will even contribute to the welfare of others, albeit indirectly. We will pass on appropriate values to our children and we will look ahead to our relationship continuing and flourishing through generations". If we would be happier with the second answer, then why should we accept for ourselves and the rest of the world the dull perspective of "just sustainable" development? Why should we not strive for rewarding, marvelous, brilliant development or, even better, just for "development", without attributes? Admittedly, in real life one should allow that in relationships, after some years, "sustainability" may become the only possible solution for the mere conservation of a family *ménage*. However, even if the main priority is the dull sustainability of the relationship, any family counselor would suggest to the partners that, in order to muddle through a sustainable relationship, they should find something new to pursue together, make new goals and eventually a new covenant between them. In this metaphor, the partners are, on one hand, humankind, society and economy; on the other hand, nature and environment, the definitions of which we will return to later.

In this chapter I elaborate on the following argument: the Sustainable Development approach has become the sole strategy available to deal with the environmental crisis and it operates as the proxy of a missing ideology. The removal of any alternative to environmental policy is paralleled with the elimination of a century old political dialectic between capitalism and socialism. This temporary lack of conflicting comprehensive political projects has impoverished the current intellectual and political debate. I will claim that the elaboration of a political alternative – based on new social and political values related to a radically new covenant between humanity and nature – would help to recreate a new dialectic and the conditions for human progress.

The sustainable development argument applies to traffic policy at large and specifically. This chapter is not essential for the explanation of my specific thesis about mobility revolution. The reader who wants to focus only on traffic policies and politics may skim through it. Although speaking about the politics of sustainable development may seem out of context, I judge it helpful because it describes the general considerations that lead me to call for a mobility revolution.

The Quest for a New Environmental Ideology

In spring 2007, with the timing and sensitivity of an experienced journalist, Thomas Friedman (2007) addressed a crucial contemporary issue in a *New York Times Magazine* article. While the 2008 Presidential campaign was entering its primary stages, he claimed that Americans did not need to choose between a libertarian or a liberal President, nor between a woman or a man, a white or a black. Rather, they needed to choose a 'green' President. The best chance America would have to play a positive role in the world would be by bringing a "new environmentalist ideology" into domestic and foreign policy. The goal of Friedman's article more than likely was to support a new Democratic nomination for Al Gore who had recently been awarded the Nobel Prize for Peace (and an Oscar Award) on the grounds of his commitment to environmental protection and specifically to combat global warming.

In principle, I could not agree more with Friedman's view. The problem with his proposal is that we cannot buy ideologies at the corner store. Ideologies develop in culture and require time to be cultivated, diffused and broadly accepted. In relation to the environmental crisis, we are just at the start of this process. In fact, we have already lost valuable time – at least 20 years – due to the unjustified enthusiasm about the sustainable development fallacy. Although Sustainable Development was nothing more than the outcome of a modest compromise following a negotiation at the U.N. Brundtland Commission in 1987, it was presented as a brilliant solution to environmental problems. Unfortunately, many activists who might have originally opposed it, eventually welcomed the compromise and failed to realize that sustainable development did not question the basics of the present economic growth model. More opportunist activists took advantage of the considerable resources that western governments appropriated to "sustain" mildly green projects. Thus, polluting industries could continue their business as usual. The mindless enthusiasm for the sustainability compromise has inhibited any intellectual and political progress toward a sound green ideology. In this sense, sustainable development has represented a conservative approach, though it has been successfully marketed and bought as an environmentalist progressive ideology.

As early as 1992, higher education institutions and western public opinion were much more advanced in constructing an environmental ethic fitting for an operative green political platform. Al Gore, campaigning for Vice-Presidency, proposed a Marshall-like environmental plan to help the former Communist countries to develop a more efficient and cleaner economy. This plan was never fully implemented, but Clinton won a number of votes, thanks to his running mate's environmental commitment. Gore's proposal was anything but a new radical green ideology; rather it was a reasonable step forward in the direction Friedman suggested.

All this happened a long time ago, in 1992. Time has gone by and now all political parties are to a certain degree concerned with environmental problems. Even the most conservative factions list environmental problems in their political agenda. To this, we need to add that we have become used to seeing and interpreting political facts through the thick lens of century-old ideologies that are hinged on the principles of liberty and justice. There is no doubt that as long as there will be two or three humans on Earth, we will have problems with liberty and justice. However, we have elaborated sound paradigms and a shared rhetorical political language to guide us in dealing with these crucial issues. The environmental crisis, on the other hand, requires a profoundly new thinking. Environmental ethics and philosophy can help in this venture once environmentalists awake from the hypnotic sleep induced by the sustainable development myth. Additionally, it is also important to recognize the risk that a true environmentalist ideology – though necessary and welcome for the safeguarding of the planet – would imply radical consequences in domestic and international politics.

The Sustainable Development Pseudo-Solution

How Did Environment Enter the Political Arena?

The environmental question entered the political arena in the early 1970s.[1] The publication of the Club of Rome's research Meadow et al. (1972), for instance, was one of the earliest influential appeals for more concern regarding environmental issues. The Club of Rome's recommendations mainly focused on the depletion of resources. At the time, the term "environment" was not as frequently used as it is now in scholarly, political and media debate, and it was certainly not employed in the same sense. Few were engaged in environmental advocacy, which was combined with other issues, e.g. Ralph Nader's pioneering consumerism in the 1960s. Environmental policy was something that was still undefined and marginal. In Western countries, the political struggle was focused on economic development and on progress in industrialization and urbanization. The competition between welfare policies and the market oriented option was at the core of politics. In countries like Italy and France, and later (after the fall of dictators) in Spain and Portugal, Communist parties, closely tied to the Soviet Union, were playing a major role in those countries' domestic policies. The environmental issue was not commonly considered by academic research and teaching. The Club of Rome itself did not dispute any other fundamental environmental issue except the possibility of running out of ores.

In the late 1970s and 1980s, a growing number of scholars opened a new research field approaching the environmental question as an ethical, political and human problem. They urged governments to adopt a specific environmental policy. Also in the scholarly milieu, several essays were published regarding the fundamentals of the environmental question in many disciplines. Economists and statisticians introduced the environmental issue in their studies by proposing new budgeting systems which included environmental values alongside the traditional financial accounting. Others proposed alternative economic systems more respectful of the environment, and called for a technological revolution and a new organization of production. Epistemologists questioned some basic paradigms that had accompanied the development of science in the last four centuries. They claimed that we needed new scientific paradigms since the old ones were the real culprits for the environmental crisis. As a matter of fact, the application of scientific knowledge to industrial production and social organization had influenced the organization of society and established powerful professional and academic guilds that hampered change.

[1] It is important to mention some groundbreaking authors such as Rachel Carson (1962) with her celebrated *Silent Spring*, not to mention Aldo Leopold (1949), among others. Nevertheless, at the time they were isolated writers, no matter how influential in environmentalist thought they have later become. However in this chapter I center my analysis on government and institutional statements and policies.

Before the publication of the Brundtland Report (U.N. World Commission on Environment and Development 1987) and the ensuing Rio Conference (1992), the intellectual atmosphere seemed ready to advance toward an alternative thinking and politics. At that time, both radical politics and thinking were inspired by a widespread perception of the environmental crisis. Political groups and environmental activists were connecting with the new cultural milieu and were trying to find new arguments to oppose the triumphant Western model. The fall of Communism was apparently imminent due to its economic, political and military failure. Western Communist parties were losing votes and in some countries they were quickly disappearing, transforming and making deals with their former opponents.

How Did We Get to Sustainable Development?

Up until the early 1980s, environmentalists only occasionally questioned the overall economic and political establishment. Both socialists/communists and liberal capitalists agreed on the idea of industrialization and technological advancement as the only possible development path. The option of a possible re-negotiation of the relation between humans and nature was marginal, if not completely foreign, to the mainstream political debate. The core of political dispute was the organization of production, namely the "progress" in exploiting the world's natural resources. Only in this respect were different options possible regarding how to distribute a growing income and reshape class relations, balance inequalities, and guarantee civil and human rights.

During the 1980s, the environmental question was becoming a very sensitive and crucial concern both for radicals and conservatives. The former could have used it as a catalyst for promoting an alternative anti-capitalist movement; the latter immediately realized the possible danger nested in the development of a possible environmentalist ideology. In international politics, the Soviet Bloc was about to collapse with the consequence of dissolving the ideology it was based on. The world's socio-political situation was favorable to creating the conditions for making the environmental discourse a credible and almost immediately available substitute for the socialist/communist ideology.

As mentioned above, regarding Thomas Friedman's call for an environmental ideology, an ideology – meant as a shared political discourse – needs a cultural milieu to flourish. It requires time to be broadly acknowledged. Then we must proceed by subjecting possible ideologies to debate in the public arena by comparing them with contrasting ones. Some 30 years ago, in environmental studies, this virtuous process – no matter how immediately and thoroughly implementable – was on the brink of taking off. The environmental crisis was widely perceived by citizens in industrialized countries, and it was becoming a regular domestic political issue. The debate would have likely driven us to the building of a new environmentalist and revolutionary ideology.

Those who picked up on environmental issues were generically considered open minded and progressive people. Environmental protection had been advanced both by radical movements and by more conservative, self-defined non-political associations and opinion-making groups, which were absolutely alien to any subversive idea.[2] As a matter of fact, environmental issues presume the necessity of a comprehensive approach and one can hardly escape being "political" when speaking about environmental policies. Therefore, the formation of new environmentalist (green) parties raised alarm among those who (a) founded their power on an ideology that negated ideology itself, (b) refused the idea of a dominant political discourse and (c) accepted the inexorableness of a society based on a neo-liberal market capitalism. It was not convenient for these conservatives to involve themselves in the discussion regarding essential considerations on ethical values or fundamental principles such as the relation between humans and nature.

Simultaneously, while the competition between the developed capitalist countries and the communist bloc was rapidly vanishing, another dualism was arising in the global economic system: China and India, preceded by some smaller South East Asian countries[3] had begun what Walter Rostow in the 1950s would have defined as the development "take off stage", i.e. a period of intense and rapid industrialization.[4] It was immediately apparent that an intense and unbridled industrialization in countries with a population of more than 2 billion inhabitants would have implied "unsustainable" environmental impacts for the planet. However, it was neither convenient nor easy for the developed countries' elites to refuse the right to development in accordance with the current technological and economic organization. It didn't matter that western citizens had begun to be seriously concerned with the health and technological risks brought about by pollution and the application of Faustian technologies, which were perceived as being too sophisticated and seemingly out of control. Western leaders needed to strike a new deal, both with their citizens and with the leaders of the booming Asian countries. Citizens from western democracies could have assumed more radical political positions, enhanced by successful green movements which were organizing themselves at the time and could have used some of the ideological apparatuses and even physical facilities left behind by the defeated communist organizations. New green parties were successfully

---

[2] An example is the conservative Sierra Club. Also the Club of Rome was anything but a revolutionary group.

[3] We refer to the so called 'Asian Tigers', namely Thailand, Malaysia and Indonesia which can only be considered "small" when compared to China and India. South East Asia countries have a population comparable to the one of the European Union, the United States and the Russian Federation.

[4] We have mentioned Walt Rostow with good reason. In the 1950s Rostow, a prominent economist and advisor to John Kennedy and Lyndon Johnson, proposed a development theory that would supposedly explain the "stages" of economic growth and was to be transformed into a development policy. His most famous essay (1960) bears as a sub title "*A non-Communist manifesto*". His ideas have been applied to South Asian development policies and have been harshly criticized, from a methodological and theoretical point of view by several scholars whose main criticism was that Rostow's development stages implied a non-political and deterministic vision. As well as in sustainable development, Rostow did not question the idea of development and proposed only one possible solution to the development process. Despite being both an influential scholar and a powerful politician in the 1960s, his simplistic ideas were soon dissolved and forgotten.

recruiting most of the radicals who were the veterans of the battles fought and (partly) lost in the late 1960s and 1970s.[5]

Nevertheless, in this period, the construction of the global market was in rapid progress and western corporations needed to reorganize their strategies and productions in order to resist the competition brought by new developing countries. If India, China and the "Asian Tigers" had not entered into the global market so powerfully, thereby opening a new quantitative growth frontier to the traditional industry, the political-industrial western apparatuses might have been more available to restructure themselves in an environment-friendly way. Moreover, in this possible scenario, the radical anti-capitalist ideology would have gone on, developing a more radical green ideology, the seeds of which had already been sown.

In academic progressive milieu, postmodern critique and the flourishing of ethical and political studies regarding the environment and the relation between humans and nature supplied the necessary cultural background for establishing a competitive political movement. To achieve this goal, it was necessary to shift the revolutionary focus from class relations to human/nature interaction. In this manner, it would have been possible either to definitely subvert the capitalist system – as a maximalist approach – or to illustrate a political alternative that reframed the missing political dialectic. I am aware of the principal difficulty in shifting the focus of the political debate from class conflict to human/nature relations. It would mean getting rid of an entire language and its tools along with its linguistic and political heritage. Nonetheless, this was – and may still be – a strategy viable to recreate a political democratic dialectic.

From Sustainable Development Compromise to Sustainable Development Fallacy

According to Dale Jamieson, "the phrase 'sustainable development' migrated from an obscure report produced by the International Union for the Conservation of Nature and Natural Resources in 1980, through several popular "green" books, to become the central organizing concept of the Brundtland Commission Report", issued in 1987. "Convened by the general Assembly of the United Nations and known officially as the World Commission on Environment and Development, the Brundtland Commission identified sustainable development as the criterion against

[5]The case of Daniel Cohn-Bendit, the leader of the 1968 Paris student revolt, is probably the most representative in Europe, but similar examples are common all over Europe. In the 1980s Cohn-Bendit became a major figure in the German Green party and still sits in the European Parliament. The same has happened to several European radical leaders who moved from former Marx-inspired communist parties (though often critical of the Soviet model) into newly founded green movements.

which human changes of the environment should be assessed, and defined it as development that 'meets the needs of the present without compromising the ability of future generations to meet their own needs'" (U.N. World Commission on Environment and Development, 1987: 43; Jamieson (1988): 183–184).

The Brundtland definition of sustainable development overtly includes the notion of "compromise". Yet the compromise was not between present and next generations, as solemnly announced. The next generation does not have a stake in the game. Thus, the real deal took place between industrialized western countries and (mainly) booming Asian economies. Actually, the word "development" has an ethical, value-loaded content, while "growth" is merely quantitative and computational. Hence, when we add an attribute to the word "development", which is meant to question and weaken the intrinsic goodness of it, we are speaking of something else rather than development. The attribute "sustainable" added to "development" implies that we need to slow down on the path along which humanity is proceeding. This means that, in the best case, we are doubtful that this development model is a promising one, or even more precisely, that this is a real development model.

The previous considerations prove that the Brundtland Commission applied a linguistic trick: they used the word "development" when they actually meant "growth", or even more bluntly "quantitative growth", as measured by customary economic indicators. Although an increasing number of western countries' citizens, politicians and scholars were once and again raising earnest doubts about the two century-old development model, the Brundtland Commission preferred to avoid a critique of it because such a critique would be refused by the developing countries, eager as they were to emulate the western affluence model. Obviously, this compromise was also "sustained" and welcomed by the conservative industrial western establishments that did not have to worry about possible drastic changes in the economic and industrial structure. If we consider the Brundtland Commission's compromise as an enlightened conservative solution, sustainable development makes a lot of sense. In fact, the international community introduced some limitations in order to slow down a deterioration process of hypertrophic growth – shrewdly defined development – that was nonetheless considered necessary to match some immediate basic needs in poor countries.

Nevertheless, how can we have a real development of humanity if its outcome is an ecological disaster which will prevent future generations from enjoying a good quality of life or even from surviving? This image is further worsened when we consider that the present generation is apparently complaining and calls for a policy change. Contemporary citizens complain about their decreasing quality of life, about the higher risk of catastrophic disasters and non-catastrophic but frequent accidents, about health disorders that include both deadly diseases (e.g. cancer) and frequently occurring illnesses (e.g. asthma and allergies). Their complaints may also stem from a personal ethical uneasiness, no matter how confused and differently expressed by diverse groups of people: there is a sense of guilt for a lifestyle that both destroys natural resources and excludes next generations from enjoying those same resources.

From Sustainable Development Fallacy to Sustainable Development Fraud

The problem is that, in the years following the publication of the U.N. Report, even those who might have been in a position to challenge the Brundtland Commission's conservative goal enthusiastically endorsed the sustainable development fallacy. Therefore, any ethical and political dialectic was swept away from the political debate. The conservative goal of the Commission was not unethical *per se*, since one may legitimately claim for both the necessity of a compromise to face immediate problems, and the need to slow down change in order to keep the system going. In the early 1990s it was still common to speak about "sustainable change" as opposed to a fast adoption of a "sustainable development" model.[6] Sustainable development was smuggled into the political debate, allowing even progressive groups to understand it as a radical change of the development model, specifically in the relation between humanity and nature. The Commission's conservative goal turned out to be unethical, or more precisely, created an unethical situation, when it became the only environmentalist credo available. World political leaders and opinion-makers succeeded in co-opting most of the possible opponents in the compromise. The "sustainable development compromise" should have been considered a transition phase toward a more radical change, but a misunderstanding occurred: sustainable development soon became the sole environmental option to challenge the development model and the two century-old relation between humanity and nature. Why did radical environmentalists immediately surrender to the successful sustainable development conservative strategy?

The real turning point was the 1992 Rio Conference that transformed the sustainable development fallacy into a sustainable development fraud. In the same year, Al Gore, who was running for Vice President in the US presidential election, published his *Earth in the Balance* (1992). Scholars, practitioners and politicians did not take his essay seriously; it looked like the customary pre-election book that candidates write for propaganda purposes, and indeed, it was that. But Gore's 1992 essay also revealed the basic environmental policy framework that was going to be implemented in the following years. Gore simplistically proposed an environmental Marshal Plan in order to help ex-Communist countries to recover from environmental disasters provoked by the overthrown regimes. The crucial idea was that you could make good business with environmental policies and that environmental protection was an opportunity rather than a limitation for development. It took a few years before Gore's proposal was unanimously acknowledged as the official credo of all governments. Unfortunately it was also endorsed by most of the oppositions and even by the most powerful environmentalist groups that in the meanwhile had flourished in Western countries. Again, *per se* there is nothing politically wrong in proposing this conservative approach. The problem was that it inhibited and neutralized

[6]The term "sustainable change" was borrowed from education and psychology literature and applied to economic change.

all the radical and possible real change which was implicit in the uprising of the environmental question in the form framed by several scholars and environmentalist groups in the 1970s and early 1980s.

To say that it was possible to approach the environmental crisis in a radical way, i.e. to choose the option of a thorough change in development policies, does not necessarily mean that this possible "revolution" was going to take place overnight, or that it ought to be directly political if not even military. Nor should we assume that it would have implied an immediate overthrow of the capitalist system. To a certain extent, the environmental revolution could have been embraced in the process of a Shumpeterian creative destruction which confirmed, rather than negated, the structure of the capitalist system.[7]

In the 1980s some authors, mainly economists who participated in development programs with powerful institutions such as the World Bank, proposed a new kind of economic development concerned with the need for a different relation with nature. These radical intellectual and political manifestoes were not desperate utopias.[8] Although few believed that they could be instantly implemented, once a possible revolutionary goal and a process were identified, there might have been several options regarding the pace of appropriate change. Prospecting a possible alternative to the business-as-usual development would have created a dialectic and creative competition in the political arena. The alleged victory of the liberal capitalist system over communism had recently removed the most popular alternative option which had been at the core of the political dialectic for 200 years. For more than a century socialism and communism, either in their revolutionary Marxist form or in more reformist and welfarist varieties, had been able to stand as an alternative to what eventually became a new triumphant ideology.[9] This missing dialectic could have been substituted by a new ideological alternative based on a different relation with nature. The sustainable development fraud prevented its plausible occurrence.

The Lost Possibility for a New Environmentalist Ideology

The possibility of establishing a green alternative to the capitalist system was lost for a number of reasons, including the sustainable development fallacy/fraud followed by the decisions of the Rio Conference. First of all, after years of opposing

[7] In Shumpeter's view, capitalism needs a repeated "creative destruction" to be able to survive and progress. These recurring destructions are anything but revolutionary as they wipe out everything except the structure of the system that remains unaffected by change. My point is that the required change might have been more environment-oriented than the sustainable development solution.

[8] Radical environmentalist authors were employed by influential institutions such as the World Bank, e.g. the economist Herman Daly who published very critical essays about the development model (Daly 1977; Daly and Cobb 1989).

[9] The new ideology, following Swyngedouw (2007: 24) is based on three unquestioned icons: (a) a neo-liberal capitalism, as an economic system; (b) parliamentary representative democracy, as the political ideal; (c) humanitarianism and inclusive cosmopolitanism as a moral foundation.

ideologies, i.e. the capitalism vs. socialism pattern, people had lost faith in ideologies and politics, hence the most successful political language was the one calling for practical solutions to everyday problems. Advocacy groups, in order to be successful and win followers had to claim to be non-political, independent from discredited party organizations and had to declare openly that they did not pursue any other long term goal except the solution of the specific problem for which they came together. People were convinced of the ineffectiveness of linking the solutions to current problems – no matter how serious and complex, like transportation policies, energy production, waste disposal – to some general change in the political organization that in addition was considered impossible. Thus the anti-system political leaders needed to be consistent with people's mood if they wanted to conserve their constituency in the short run. They were confused and drifted because they were forced to give up their ideological toolbox and substitute their way of thinking and acting. At the same time, the construction of an environmentalist ideology required time and no substitute for the old schemes was yet available.

There were many factors playing in favor of a possible opposition based on environmentalist politics. Among them, there was (a) the mounting fear of technological risks and man-made natural disasters which had produced a rich literature on risk as a characteristic of contemporary society (Beck 1997); (b) the growing inefficiency of the solutions to urban problems offered by customary technologies; (c) the crisis of science, of the scientific method, and of the trust in the scientists' credibility (Latour 2004, 2005); (d) an emotional rejection of the artificialization of the world, brought on by extensive and untamable industrial production. Capitalism – in its new forms – had been successful in guaranteeing affluence to many people, mainly in the Western countries. It was easy to think that the same system could work effectively also in the less developed countries, possibly with some minor adjustments. The income differences between poor and affluent countries were so striking that, in the developing countries, people's hope for the future more than compensated for the perception of problems connected with unequal distribution and with environmental risks.

On the contrary, citizens of western countries no longer hoped for further improvements in the traditional way of life, and, as a result, the elaboration of a new relation with nature was becoming an option. Along with this possible new relation with production and nature, it would also have been likely to propose a new model of citizenship which was able to create a new association between people and places.[10] It might have included innovative considerations about area and administration, federalism, self-government of communities, welfare states and so on. This would have helped in integrating global migrants by applying more updated paradigms rather than those developed for no-longer existing national states. It was possible to claim that the system was unable to coordinate with nature and was economically inefficient

[10]The bio-region movement tried to say something *à propos*, but the literature on "bio-regionalism" has never taken the political question seriously, thus remaining completely confined to geographical and biological studies.

if non-monetary values were taken into consideration. Finally, the widespread distrust of ideologies and grand discourses, the lack of a political leadership interested and educated in environmentalism, and eventually the success of the sustainable development fallacy/fraud, hampered the creation of an alternative.

To create consensus around the sustainable development industrial policy, governments needed to defuse the oppositional power of radical environmentalist political parties and movements that aggregated in the 1980s. Because of people's strong sensitivity to the environmental crisis and the consequent success of green parties and the like, these groups had become a significant stakeholder in the political debate and might have opposed this conservative policy. The sustainable development fallacy/fraud was applied also to political groups and leaders. Most of the leaders of the green parties and environmentalist advocacy groups had been educated in a political philosophy almost completely unaware of the environmental issues and ethics. Their political training and experience was based on civil rights, labor relations and equality issues. Only some of these issues had an environmental content and generally environmental concerns occurred as a tangent to other concerns such as in reducing cancer risks for workers employed in chemical industries. Due to the growing environmental awareness of citizens, major parties supported parallel environmentalist associations. These associations were meant to operate outside official politics, which were more and more discredited in coping with environmental decision-making. At the same time, political leaders of major labor parties were able to keep environmental discontent under control through these associations. Some of them, namely the ones closely connected to the parties, became very powerful and large. Thus they had to hire staff and manage thousands of volunteers whose organization eventually needed to be somehow financially "sustained". When it was not possible to integrate the most radical organizations into the SD policy, they were considered outside the law, and treated as extremists, if not outright terrorists (Agamben 2005).

The political discourse did not entail the specificity of the environmental question and relinquished the crisis as much as possible into the hands of scientists. Notwithstanding the growing mistrust for science and scientific corporations, people expected a solution from rational and possibly trustworthy scientists, rather than from a rational ethical discourse (Shrader-Frechette and McCoy 1993). This attitude helps in avoiding the questioning of the power of scientific and professional corporations, no matter how crucial it is.

A Bike Route Beside the Nuclear Plant

The Rio 1992 Conference was the turning point for the definitive defeat of environmentalism as a possible alternative to the capitalist system. The idea that governments should invest in environment-friendly technologies and behaviors was welcomed by everyone. Governments took advantage of the new situation by funding industries in order to allow them to reconvert a part of their productions and make them "sustainable." Governments' funding did not really modify the staple productions and the overall organization. It allowed new investments in lower impact technologies, whose real effectiveness has been often questioned.[11]

The disposal of a possible environmental revolution was accomplished. After the 1992 Rio Conference,[12] all developed countries' governments financed sustainable development programs and projects that "sustained" a growing number of self-defined 'environmentalist' associations. Research in sustainable development industrial technology was also copiously sponsored. The Marshal Plan idea was *de facto* implemented although it primarily benefitted the ecological reconversion of western economies toward a marginally more friendly relation with the environment rather than being applied to developing countries that have gone on adopting older technologies with higher and higher environmental impacts.

The paradox was that often, in order to produce lower environment impacting final products for developed countries, most of the ecological components were manufactured in countries with no environmental regulation applied to the production process. The funds granted to environmentalist advocacy associations for implementing marginal projects in the name of sustainable development were crucial in the manipulation of public opinion that eventually assumed the Brundtland Commission's approach and the Rio Summit's policy tools as a conclusive solution. Former advocates of an environmental revolution lost interest in acting politically, as they were: (a) busy applying for funds available for ecological projects, (b) involved in cooperative efforts with industry, (c) focused on implementing minor local projects, and (d) lacking a coherent ecological system of values.[13] They were content with the smaller, more visible successes such as, for example, they showed pride in having realized a bike path in a neighborhood, despite the fact that it was located near a nuclear power plant.

Environmentalism as a political ethical philosophy lost a large part of its influence and became unable to produce any real change. Environmental policy remained firmly in the hands of those who had created the problems, i.e. scientists, industrialists and technicians.

[11] A good example of this is the car industry that in the sustainable development era did not make any significant progress in reducing emissions although it has regularly obtained relevant incentives. Moreover, although the emission problem has been taken somehow into consideration, very little has been said about all the other environmental impacts generated by traffic, car construction and recycling, parking, road construction, social life, etc. The sustainable development fallacy/fraud also introduced the sanctification of public transportation which justified the construction of transportation infrastructures that added to total mobility, and (consequently) impacts, rather than operating as a substitute of private cars or, even better, for reducing the overall mobility. The sustainable development allowed the building of new waste incinerators, supposedly clean, instead of questioning the disproportionate production of waste. And so on.

[12] Another landmark of the sustainable development implementation policy was the European Conference of Sustainable Cities and Towns, held in Aalborg (Denmark) in 1994.

[13] Except for point (d) all these outcomes of sustainable development strategy have had some positive effects: e.g. they have created an environmental consciousness among the people who are now more informed and educated in environmental issues.

How to Re-Frame the Environmental Political Case

An Epochal Divide

Since antiquity, philosophers have argued about the relation between humanity and nature. However, around the middle of the twentieth century the relation changed more than it had ever done since humans appeared on Earth. Hannah Arendt proposed that a symbolic date should be fixed on the day when Yuri Gagarin was the first man to see the earth without being part of it. That epochal date could also be fixed on the day Hiroshima and Nagasaki were bombed. In that moment, for the first time in human history, it became clear that nature should fear humanity. Until then humans were awed by nature, as it was mostly beyond their control. This epochal juncture could also be said to have been reached the moment the human genome was mapped, opening the way to cloning the human being and to the biogenetic revolutions already in progress. All this happened in the mid 1900s. Arendt, like other philosophers and epistemologists, identifies the remote origin of this epochal change in the diffusion of the Copernican theory and in Galileo's discoveries, which proved that:

> the worst fear and the most presumptuous hope of human speculation (…) and the Archimedean wish for a point outside the earth, from which to unhinge the world, could only come true together, as though the wish would be granted only provided that we lost reality and the fear was to be consummated only if compensated by the acquisition of supramundane powers. For whatever we do today in physics, (…) we always handle nature from a point in the universe outside the earth. (Arendt 1998: 262).

The radical epistemic critique has been a major source of inspiration for many environmentalist scholars. This became widespread in the 1970s, led by Berkeley's Thomas Kuhn, Imre Lakatos and Paul Feyerabend whose thinking is also related to Adorno and Marcuse's critical views of contemporary society. Probably the most elegant description of the new situation was offered by Hans Jonas' reading of Sophocles' Antigone choir.[14]

> Wonders are many, and none is more wonderful than man; the power that crosses the white sea, driven by the stormy south-wind, making a path under surges that threaten to engulf

---

[14](http://classics.mit.edu/Sophocles/antigone.html, translation by R. C. Jebb). Jonas is mainly quoted for his "imperative of responsibility", i.e. the responsibility that contemporaries should have toward future generations (Jonas 1979). In this respect, Jonas is usually included among the so-called "anthropocentric" environmental philosophers (who are opposed to "eco-centric" ones, i.e. the ones who claim a parity between human and nature rights) because he builds his ethics on human needs. However, there is another crucial theme of Jonas' arguing that is sometimes overlooked, i.e. the consequent responsibility for nature on which human future generations depends. Thus, what really matters for Jonas and for the "imperative of responsibility" is neither humanity nor nature, but the relation between the two as implicitly reported in Jonas' opening quotation of Antigone's choir. Jonas, moving from the crisis of Western rationality, prospects a radical change in the subject/object relation with respect to nature and humanity, to the extent that he can be legitimately included among the so-called "deep ecologists" (Tallacchini 1996: 4).

> him; and Earth, the eldest of the gods, the immortal, the unwearied, doth he wear, turning the soil with the offspring of horses, as the ploughs go to and from year to year.
>
> And the light-hearted race of birds, and the tribes of savage beasts, and the sea-brood of the deep, he snares in the meshes of his woven toils, he leads captive, man excellent in wit. And he masters by his arts the beast whose lair is in the wilds, who roams the hills; he tames the horse of shaggy mane, he puts the yoke upon its neck, he tames the tireless mountain bull.
>
> And speech, and wind-swift thought, and all the moods that mould a state, hath he taught himself; and how to flee the arrows of the frost, when 'tis hard lodging under the clear sky, and the arrows of the rushing rain; yea, he hath resource for all; without resource he meets nothing that must come: only against Death shall he call for aid in vain; but from baffling maladies he hath devised escapes.

Jonas, in the beginning of his essay (1979), maintains that these lines no longer describe the current relation between humans and nature. For the purpose of these introductory notes, we need to add and focus also on the following lines of Sophocles' Antigone Chorus:

> Cunning beyond fancy's dream is the fertile skill which brings him, now to evil, now to good. When he honors the laws of the land, and that justice which he hath sworn by the gods to uphold, proudly stands his city: no city hath he who, for his rashness, dwells with sin. Never may he share my hearth, never think my thoughts, who doth these things!

This last part of the chorus is meant to restore the question of the human/nature relationship to a political and ethical discussion.

The Structure of an Environmental Problem as a Political One: Four Stages of Consciousness

Before approaching the critique of sustainable development pseudo-ideology, firstly we need to define the environmental problem from the political point of view.

Analyzing the definition of the environmental problem from the political perspective in the contemporary environmental political debate, we may identify four types of environmental consciousness which give way to consequential political actions. We may assume that the four types are also stages, ordered in terms of intensity of concern for the environment. Moreover, although the four types are neither listed nor intended as a chronological succession, I stand for an affirmative evolution from the first to the fourth.[15]

At the first stage, we can completely deny the existence of problems defined and catalogued as "environmental". The "environmental file" comprises so many entries that it has become predictably nonspecific. Air, water, and noise pollution, as well as waste disposal, traffic congestion, endangered animal and vegetal species and so on, are such diverse phenomena that they are not necessarily supposed to be grouped together. Different disciplines and diverse professionals are in charge to study and deal with the aforementioned list of problems. Obviously, this position is now

[15] I first proposed this classification in Poli (1994a: 125–141).

outdated: at least three decades ago we learnt to catalogue a group of issues defined by common sense as 'environmental'. However, although we have adopted a different taxonomy and have grouped them in the novel entry named "environmental", when we deal with the problems in practice, we still approach them separately. Until a quarter of a century ago, most scholars still refused the idea that a comprehensive approach to environmental problems was necessary. From an epistemic point of view the prevailing idea was that the progress of each science and the advance of applied technologies used by professionals and practitioners was the obvious solution to problems that were not "environmental" but rather chemical, biological, physical, engineering, genetic and so on. Until 1980, there were very few higher education programs in anything called environmental studies, nor had traditional teaching subjects – such as chemistry, engineering, geography – yet added the adjective "environmental" to indicate either a new content or an innovative approach. Economics was the discipline that would have synthesized, in the market monetary solution, each single problem that was worth separating from the others.

The second stage entails a higher degree of concern regarding the environmental crisis: people admit that the relationship between humans and the environment ought to be somehow revised. Therefore, we select a series of different issues worth being categorized in the same entry as "environmental". This is a step forward from the first stage because the new classification is meant to lead to building new links between phenomena and situations, and focusing on these links rather than on the single issues. From the scientific method point of view, we can tentatively assume that the first stage accepts the classical reductionist approach; while this second stage is more concerned with an interdisciplinary approach and/or with system analysis.[16] Nonetheless, the scientific approach is still prevailing over any political and philosophical arguing. Thus, the proposed solutions to the newly grouped-together problems, recently defined as "environmental", proceed in the traditional fashion, i.e. keeping them rigorously separated when we need to manage them practically. Nowadays, this is the most broadly adopted approach. It implies that all environmental problems can be addressed in a purely technical manner. The approach insists that to solve environmental problems, it is enough for each operative organization to merely contribute to their own part. This consequentially leads to a comprehensive vision. At the most we can speak of interdisciplinary knowledge and coordination. Sustainable development belongs to this level of concern. People endorse this second level when they assume that traditional technological progress is not only the sole possibility to solve environmental problems, but that this type of solution will also favor further economic growth.

At the third stage, we break into the field of real environmentalism,[17] which can be more or less extreme. In other words, we can avow a shift from environmental "issues" into an environmental "question". This is something substantially different from the

[16] For this reason there has been intense debate among epistemologists if "ecology" should be considered a "subversive science" (Shephard and McKinley 1969). The ethical and epistemic debate about science and environment, still quite alive in the '1990s, focused also on how to use ecological methods to understand and/or handle environmental problems (Shrader-Frechette and McCoy 1993).

[17] The suffixes "ism" and "ist", added to the word "environmental", indicate the idea of development of a political movement which goes beyond the single issue and entails a social critique.

basic recognition that there exist problems to do with the environment, which can be conventionally grouped and possibly given priority over other problems. This third level of concern implies the existence of a real "environmental question" whose solution would require a change in lifestyle, ethics, laws, technology and production systems. New techniques are not enough to solve the environmental crisis. Rather, we need a new scientific approach, new paradigms. From this perspective, the environmental crisis is not a technical problem. Instead, it is an ethical, social, organizational, and ultimately, a political issue. For this very reason, environmentalists insist on actions unrelated to traditionally fragmented bureaucratic competence. They move beyond solutions organized around the operative and administrative structures of most governments. Consequently environmentalism is revolutionary, i.e., in order to overcome the typically conformist responses to environmental problems, environmentalists believe that the problem should be approached mainly from a political point of view. They assert that, if we do not intend to change the existing relationship among humans, technology, and nature, then the current breakdown is just being "patched up" using known technology. In this case, current social, political and scientific structures are preserved, becoming even more powerful and sophisticated.

The approach outlined as second stage is not "environmentalist." It demonstrates a generic sensitivity to environmental problems that can be somehow shared by virtually anybody, independent of their political beliefs. Environmentalism is meant to pick up the environmental issue as a political one and associate it with other crucial political issues such as labor relations, civil rights, citizenship, political participation, citizens' privacy, tax systems, etc. Moreover, the environmental question has its own specificity and is different from other more traditional political questions.

The fourth stage of concern is the most intense. Radical environmentalists claim that the environmental question today is the pivotal political issue around which all other political and social problems orbit. Environmentalism is viewed as the approach to start with, in order to solve all other political and social issues. In the last three centuries of human history, the political debate has been hinged on social justice and individual freedom. Most of the political theory elaborated in this period was conceived in relation to different and contrasting ideas on how to combine and pursue social justice and individual freedom. This was happening in an era in which a growing wealth needed to be redistributed among peoples and social classes. All other considerations were often deemed a consequence of this priority.

Radical environmentalists would claim that we need to start our political militancy and our theoretical elaboration with considerations about a new deal between humans and nature. Although social justice and individual freedom will always be crucial, the starting point of the political debate should give priority to environmental preservation, non-human entities rights, the relation between people and territory, bio-citizenship, etc.

The two intermediate positions (stage two and three) are the most likely to be adopted. Yet, the political success of the U.N.'s sustainable development approach has overshadowed the third level. These two positions – both seemingly reasonable and moderate – are indeed separated by a clear philosophical divide which involves opposing environmental and political ethics, and a non-reconcilable epistemology. The dramatic break between these two positions has been underestimated and overlooked. As a result, we do not want to try a negotiation between these two philosophically distinct and seemingly incompatible positions. Rather, we need to make the conflict discernible and hence make both these positions "political", instead of leaving them to rot in a sterile academic controversy. As we will see below, a new dialectic – which substitutes the vanished dualism between Western democracies capitalism and Soviet communism – can be created if the crucial difference is recognized and given political status.

In the following sections I will try to prove how an authentic and radical environmentalist way of thinking requires citizens to intervene differently in the political controversy. We must end the idle focus on piecemeal solutions. My position is that the only chance for change is through a cultural revolution which locates the environment at the center of the political debate and makes all other issues subsidiary. Some increasingly unbearable situations, regarding the deterioration of both physical and social conditions of life, may be the fuse to prime the process, i.e. a favorable starting point for creating a political alternative to neo-liberal capitalism.[18]

Definitions of Nature and Environment: Impossible 'Sustainability' and the Post-Political Condition

Nature and environment are very different concepts and there is remarkable philosophical arguing about their meaning. Dale Jamieson has claimed that in some cases they can be used indifferently. Although I agree with Jamieson in relation to the theme of the book in which he advanced this utterance (Jamieson 2008: 2), I argue that this distinction between the two concepts proves useful when we introduce the problem of sustainable development's pseudo ideology.

The word "environment" comes from French and means "all that is (turning) around you". It concerns both human built artifacts and that which interacts with humans. The use of this word has recently become popular also in scientific and economic language, e.g. in ecology. From an epistemic point of view, using the word 'environment' suggests a shift from a science focused on the particular, to the analysis of the relations occurring among diverse and multiple phenomena. Instead, the word and the concept of "nature" are as ancient as philosophy itself and probably older.

In an inspirational essay, Erik Swyngedouw (2007) elaborates on Žižek's argument that there are "several natures" and that the construction of nature is a political action (Zižek 2002). Although this statement may be accepted in principle, nonetheless, if we neglect to distinguish between nature and environment, we lose a critical tool to readdress environmental policies. In fact, by moving from the second to the third stage of environmental concern, as previously described, we may find it helpful to distinguish between environment and nature.

[18] In a certain sense, you do not have to be necessarily "anti-capitalist" or thoroughly revolutionary to support the creation of a sound alternative to neo liberal capitalism. Facing an opponent may even help to reinforce the current system.

On one hand we accept to apply the term "environment" to the product of different cultural-political constructions, according to Swyngedouw and Žižek. On the other hand, we also use the concept of "nature" to describe an entity or subject that holds its own standings although it is not part of humanity.[19] In other words, we introduce another "subject" different from humanity rather than assuming that we deal with nature as with an "object" or even a human artifact. Hence, we assume that a singular nature – which, as a subject, is intended as an end *per se* – should be considered in the environmental*ist*[20] political debate. At the same time, we can maintain that it is possible to conceive and design multiple environments. In the first two stages, we do not need to distinguish between environment and nature, the latter being an object owned by humans and possibly within their control. Nature is completely reified. That is why environmental scientists and activists "invariably invoke the global physical processes" and "insist on the need to re-engineer nature so that it can return to a 'sustainable' path" (Swyngedouw 2007: 20). The questions are: (a) Do we hold the right to treat nature as an object and, hence, are we allowed to engineer it without any moral limit? (b) Do we consider humanity as part of nature thus humanity should be morally committed to her conservation as much as to its own? If we respond affirmatively to the first question, we do not need to distinguish between nature and environment. If we respond affirmatively to the second question, then we need to consider human development intrinsically connected to and dependent on the respect for nature. The choice here is between exploitation of nature and attunement with nature. Both positions may imply moral limits to growth and both may presume a transformation of nature. However, the relation between humans and nature, the consequent policies, and the conditions for the construction of an environmental discourse are very different in the two cases.

The idea that environment is not distinguished from nature and that both environment and nature can be engineered and handled with economic and technological tools, is the keystone of sustainable development policy. The human and political dimension of the environmental question has been circumscribed into the borders of well-established economic and technological paradigms elaborated in order to embrace, manage and understand the functioning of society. What has been ignored is the existence of nature as a separate entity. As long as we consider nature (and hence "natures", as Swyngedouw/Žižek claim) as a projected image of humanity and fail to give it a subjective status, we ignore the necessity to renegotiate human/nature relations on an ethical and thus political ground. As a consequence, we also miss the opportunity to radically criticize the production system and the political-geographical

[19]There is an abundant bibliography about the dualism between antropo-centric and eco-centric environmental ethics and about the distinction between "Deep" and "Shallow" ecology , to quote Arno Naess' original words. A pioneering essay on the theme bearing an effective title is Stone's (1974) "*Should Trees Have Standings? Toward Legal Rights for Natural Objects*".

[20]I distinguish between "environment" activists and scientists, from "environmentalists " according to the previous classification.

content of citizenship – i.e. relations between the sustainment[21] of peoples, settlement, and political organizations – from this point of view.

Swyngedouw and Žižek do not discuss the definition or the status of nature. In the sustainable development approach, it is not nature that we consider harmonious: what is harmonious and unquestionable is the production system, the market and political economy which is assumed as (or very close to) a natural science. In this respect Swyngedouw and Žižek's position is much closer to the sustainable development approach than it may appear at first glance. Their critique explores the same battleground where sustainable development maneuvers. sustainable development supporters include nature in economics, depriving it of any political and ethical status. Therefore, as Jamieson puts it (2008: 22), "disagreement is allowed, but only with respect to the choice of technologies, the mix of organizational fixes, the details of the managerial adjustments, and the urgency of the timing and implementation". If this is correct, it applies exactly in the same way as the Marxist approach, which implies that social change is fundamentally driven by economic facts and that "environmental problems are caused by the distribution of property rights and incentives". Sustainable development economists and Marxists may disagree about "exactly what is the correct explanation, but they agree about the terms" (Jamieson 2008: 22). For both of them, the correct explanation of environmental degradation is one that is fundamentally economic in character. This position is justified if we assume a materialistic idea of nature and believe that humans are driven by natural forces that determine our behaviors. In a situation like this there is little place for deliberate rational human political action. Although it is acceptable to define this situation as "post-political" (Žižek 1999: 35, 2006; Mouffe 2005), as reported by Swyngedouw (2007: 23), and because we have accepted to critically define this situation as post-political, we need to go beyond this dead ended approach and look for what can become "political", what we can transform into "political" in the coming years.[22]

Swyngedouw is convincing when he describes the post-political condition built around the inevitability of "neoliberal capitalism as an economic system, parliamentary

[21] Another ambiguity of sustainable development concerns another meaning of the word "sustainable" that recalls the necessity – crucial for poor countries – to provide an adequate "sustainment" to very poor, mainly African, countries' people whose economy has been devastated by globalization and urbanization. Sustainable development may be intended also from the perspective of "sustaining" (feeding) starving peoples. In this perspective, long term environmental sustainability becomes secondary to the priority to provide food to people. Thus, another possible limit to growth is canceled on the ground of a moral principle that helps in avoiding any discussion on the capability of the capitalist system to both create and solve regional poverty.

[22] An everyday life consequence of this way of intending sustainable environmental policies and social change is that, when we try to organize an environmentalist advocacy committee in any town of the Western world and want to win followers, we need to reassure potential members that we do not have any political intention, not to mention any party affiliation. All that we are allowed to declare is that we want to solve the specific problem. Then we waste time craving a scientific explanation on which nobody will ever agree.

democracy as the political ideal, and humanitarianism and inclusive cosmopolitanism as a moral foundation". However the next step should be to make an effort to move from the post-political condition to a pre-ideological or neo-political condition. True that this is not an easy task, but, as Žižek (quoted by Swyngedouw) proclaims, "authentic politics … is the art of the *impossible*" (Swyngedouw 2007, emphasis in Žižek's original) or as a famous 1968 motto goes: "be realistic, go for the impossible".[23]

Sustainable Development and the Science Authority

The weak point of contemporary populism, of which sustainable development is one of the most prominent tools, is the necessity to rely on legitimized science and technocracy. But, as Sheila Jasanoff (2010: 695) has recently written, "science and technology are rapidly losing their reputation as the only possible saviors of humanity from natural danger and poverty. It is no longer enough to establish what counts as good science; it is equally important to address what science is good for and whom it benefits". On one hand, science has become the slave of a technology dominated by economic and professional corporations, so much so that it has become difficult to rely on the independence of scientists and their institutions. On the other hand, the complexity of the effects of scientific and technological innovation has made it necessary to introduce value and ethical considerations, regarding technology and research, in any assessment. In the populist system, citizens desperately quest for an independent scientific opinion which should come from a legitimate scientist in order to permanently solve problems. However, the truth they long for is not going to emerge and instead, the decision comes from a process of negotiation. Negotiation would be political *per se,* except that the participants involved deny the political content of the decision. The consequence of this refusal to consider the intrinsic ethical-political content of the decision is that nobody focuses on devising political institutions suitable for dealing with environmental decision-making which may include ethical and political arguments as well as scientific and technological matters.

Swyngedouw points out the populist tactics of "not identifying a privileged subject of change (like the proletariat for Marx, women for feminists, or the "creative class" for competitive capitalism)" (2007: 33). The proletarian revolution seems to have failed, although the dialectic juxtaposition between capitalism and communism and the labor movement have helped to enhance workers' rights considerably.

[23]"Brilliant politicians are not the ones who keep promises made to their constituency. If politicians maintain all that they have promised, it means that they have not promised enough". This is what I proclaimed while running for office a few years ago, and probably was the reason why I failed … in this mediocre world. Yet, I still believe it, although I am not going to utter it anymore during a campaign, in the unlikely case I will run again for office. Innovative politicians must propose a vision and show the way.

The traditional proletariat is no longer a possible subject of change. Feminists and women are a more viable subject of change *vis à vis* the environmental problem since feminist theory has already argued that the women/nature relation is radically diverse from the dominant men/nature exploitative and "male-rational" relation (Plumwood 1991). I believe that this approach is helpful, but it can only really work as a side argument. In fact, it is divisive and it is not likely to become a shared discourse. The so-called creative class, proposed by Richard Florida (2002), also mentioned by Swyngedouw, operates inside the current system, follows business as usual, and adopts a predictable technology. Thus we assume that creative class, now so popular, is nothing more than a conservative approach and it cooperates to the sustainable development fraud.

This gloomy scenario can be made brighter if we try harder to identify new possible subjects of change. Global warming has been fetishized and used as a bogey to justify an international policy to allegedly protect the entire environment. But people perceive the effects of the environmental crisis in everyday life in different and diverse areas: the amassing of waste, the loss of green land, noise, disproportionate use of energy, traffic, growing distance between place of production and place of consumption, etc. These everyday problems have intense psychological effects since they create an overall sense of up-rootedness. This crisis is also to do with the 'built' environment,[24] such as urban design, grey shopping centers and their never-ending parking lots, city slums and mass housing in metropolitan peripheries. All this produces a lifestyle that is very dependent on driving, living in isolated houses and apartments, shopping in anonymous global grocery stores, packaging and preserving food, commuting to work and to any other daily activities including leisure. Some claim that this lifestyle is the outcome of people's market choices. It is true that we choose this lifestyle, but only because there is no alternative available for most of us.

These feelings of refusal of a noisy, dirty, isolated and high environmentally impacting lifestyle – i.e. a rejection of the most common organization of contemporary life – have spread among several individuals and have become a conscious attitude for many. The sociologist Paul H. Ray and the psychologist Sherry Ruth Anderson (2000) claim to have identified 50 million adult Americans and another 80–90 million Europeans who they define as "cultural creatives". These people would be willing to change their standard lifestyle, albeit with different intensities, if they were given the chance. Even if one may dispute the details of Ray and Anderson's research, the very fact that they have tried to identify a subject of change means that there is a quest and a need for it. The problem is if and how it is possible to transform the "cultural creatives" from a sociological classification into a political subject of change.

[24] Warwick Fox (2000) in his introduction to "*Ethics and the Built Environment*" (a collection of essays) claims that we should pay more attention to urban environment since nowadays most people live in cities, the organization of which is responsible for human health, quality of life and risks.

A New Dualism Needed

We should not overlook the fact that the cosmopolitan order may be fragile and short term. We can already envisage an emerging political and ethical dualism between the West and Islam. Hence the so-called cosmopolitan order really works in Western countries, but it is harshly questioned and fought on a global scale. In the last decade Islam has become a political adversarial theory which embraces some modern beliefs and rejects others. Islam fights against capitalism and secularity opposing religious and political principles. On one hand, this is still an old fashioned, non-global approach to world politics since it takes into consideration religious, ethnic and territorial conflicts. On the other hand, Islam has spread all over Europe and North America and proliferates in the still hegemonic cosmopolitan order. In Europe and North America Islamic organizations are flourishing and they often catalyze – better than traditional leftist-socialist parties – the poor's discontent giving them hope and identity: a cause to fight for. Because of the global order, the most likely scenario is a dialectical conflict that will take place inside the global society. If it is advanced by violent ideologies, it may explode into civil wars and into a series of riots and never-ending turmoil. In Europe and North America, the anti-Islam parties and ideologies, that are becoming popular and politically meaningful, reinforce the anti-western, anti-modern Islamic movements rather than curb them.[25] In this context, the goal is to change the system and found a new political, democratic and just order based on values different from most of the ones put forward by anti-western Islam political-religious movements. Old-modern western society's likely reaction is to fight violence with violence with little possibility of success.

Instead, a radical substitute for West-Islam dualism's perspective, which is a dialectic alternative to the postdemocratic, postpolitical, cosmopolitan order, would consist in an effort to construct an intellectual and hence political new ideology based on a different relation with nature. This is nothing new as it would mean reconnecting with environmentalist movements whose origins go back to the pre-sustainable development green movements. These movements – in which we may include the New Age pop-culture[26] – and the theoretical thinking that supported them, have never disappeared although they have temporarily lost their appeal because of the sustainable development fraud. Modernization and capitalism have been extensively studied from a social and political perspective. Several scholars have also addressed the dramatic changes that industrialization has provoked on

[25]It has been questioned by many contemporary authors if Islamic integralism is indeed a post modern feature rather than an attempt to restore a traditional order.

[26]The New Age movement is meaningful because it developed in the West but was characterized by its heavy borrowing from native, eastern, Buddhist, and in general "other" cultures.

the relations between society and nature. However, the latter has never gained the limelight in political discourse.[27]

One could paradoxically argue that some natural or human-provoked catastrophe could help to regenerate the human species and therefore we should not concern ourselves so much about the present situation. This absurdity raises the point that most of us emotionally and rationally refuse the idea that a human catastrophe may be worth regenerating and reinforcing the human species on the ground of an ethical judgment. Likewise, why should we not accept that restoring a fair balance between humanity and nature is a moral principle we should endorse? As Mouffe (quoted by Swyngedouw) rightly claims, bio-engineering, extremely advanced de-humanized medicine and bioethics have boisterously entered the political arena creating new divisions among traditional political and religious parties and opinion groups. Thus, why shouldn't a more extended concept of bioethics, applied to the overall relation between humanity and nature, become a crucial political battleground? In the eighties many scholars and research projects considered environmental ethics and bioethics as part of the same discipline or study area (Tallacchini 2009; Stevens 2000). As for the New Age movement, this new approach would have the great advantage to be born in the Western culture, but with the main intent of borrowing and integrating with other non-Western traditions. Therefore, differently from the Islamic antagonism, a radical environmentalist opposition to the neo-liberal capitalist order, would be deprived of a territorial and traditional content and would become purely ideological.

We would come to a dead-end if we grounded our search for a new policy to contrast the current situation, on the usual class exploitation and social justice categories. Historically, one of the main characteristics of capitalism and industrialization[28] has been high energy consumption and resources transformation. This metabolic aspect of capitalism, although it has been treated by scholars since the beginning of the industrial era, has always been kept aside in the political debate. Even the issue of body politics has not been passionately dealt with in the natural realm. For a long time communist and socialist parties have considered the industrial revolution as a "chicken of the golden eggs"[29] since industry was seen as a tool

[27]The opportunity to readdress political and psychological discourse in order to make it more suitable to deal with the new relation between humanity and nature was developed in the early 1990s by the political philosopher Eckersley (1992) and by the ethicist and psychologist Fox (1990). They identified the eco-philosophical attitude as identification with others and as an extension of the self. This approach introduces eco-feminism that tends to associate women and nature as they are both the object of men's exploitation (Tallacchini 1996: 58–59). More recently, the traditional feminist Badinter (2010), in a very controversial pamphlet, has argued that ecologism is an enemy of women. Eckersley and Fox are just two of the several scholarly attempts to create a new political and ethical basis for a revolutionary environmentalism advanced in the early 1990s.

[28]Paul Samuelson provocatively claimed that the Soviet Union production system was not less "capitalistic" than western countries'. In fact, the Soviet industrial system devoted as much physical capital as the so-called capitalist economies to production, except that it was state owned capital.

[29]This is the definition of capitalism given by Filippo Turati the founder of the Italian (reformist) Socialist Party at the turn of the nineteenth and twentieth centuries when the Northern Italy industrial revolution was taking place.

for the *liberation* of *humanity*[30] from mass poverty. Therefore socialist parties, that were concerned with exploitation and redistribution of the new industrial affluence, did not discuss, and therefore actually preserved and encouraged, the transformation of the economy that was taking place.

Hence, I agree with Swyngedouw's opinion that environmental populism – evoked by the threat of an impending environmental catastrophe – "silences ideological and other constructive social differences, and papers over conflicts of interests by distilling a common threat or challenge to both nature and humanity" (2007: 32). But, in order to move the debate further, we should conduct research to answer the following questions: why do people rely on populism and disregard other solutions? Can we simply utter that people have been effectively manipulated by the sustainable development fraud? Can we just blame capitalism because a shift occurred from political to post-political condition? If the current situation proves that a new post-political ideology has taken over the old communism *vs.* capitalism dualism, and "all people", at least in the western affluent countries, "are affected by environmental problems", why should we apply the same analytical and ideological concepts that have been used in the past when the main social problem at the time was income distribution? It is not effective to propose the same approach to a new situation; there is the need to find a new one. Finally, we should consider that the present post-political condition is not the outcome of a unilateral decision of a presumed ruling class, but it is itself the result of a conflict. The environmental issue has been put forward by the opponents of the capitalist system, who chose a new battleground instead of the traditional class conflict for the redistribution of growing production.

Conclusion

Hence, I agree with Swyngedouw´s analysis and Slavoj Žižek's belief who, in *Looking Awry*, suggests that the current ecological crisis is indeed a radical condition that not only constitutes a real and present danger, but, equally importantly, "questions our most unquestionable presuppositions, the very horizon of our meaning, our everyday understanding of nature as a regular, rhythmic process". This crucial utterance clearly paves the way for a broader argument about the centrality of the environmental question in the political debate. The clean break that occurred in the twentieth century has overthrown the idea of a nature too powerful to be endangered by human action.

Žižek and Swyngedouw claim that it is easier to predict an environmental catastrophe than to envisage relatively small changes in the socio-political and cultural-economic organization of local and global life here and now. It is definitely easier and customary to foresee the final ecological Armageddon rather than a transformation of

[30]Italics because: (a) "liberation" is intended not only from social class exploitation and injustice as the extremist socialists were calling for, but also from contingent poverty; (b) only "humanity" matters and no consideration is given to any kind of interspecies or natural justice which, at the time, was an issue completely ignored by the political debate.

the neoliberal capitalist order. But why is this happening and how can we change this condition through political action and intellectual elaboration? *Prima facie*, we can argue that a millenarian attitude can either work for conservation or for change. Therefore, we need to answer another question: why do contemporary people – mainly western affluent countries' citizens – fear a possible Armageddon to the extent that it has secured such a noteworthy position in contemporary political discourse? Why is this genuine widespread fear, founded among people, causing such a strong reaction, to the extent that the environmental crisis, much more extensively than any other political emergency, has been chosen by the conservation forces, i.e. the neoliberal capitalists, as the main battleground, and the cornerstone of their customary and cyclical "revolutions", and is skillfully managed in order to preserve the system?

Do we really need the threat of an Armageddon in disguise of global warming or nuclear power to promote social change? Is it so impossible to simply long and fight for an improvement in our lives, i.e. in a real development rather than a depressing "sustainable" one?

We should accept that the environmental crisis is different from all the previous ones. First, as reported above, an epochal change, never experienced before, occurred in the mid- twentieth century. As a consequence, the environmental crisis should not be analyzed as one of the recurrent revolts against an unjust political system and should be managed in the usual way. As a matter of fact, sustainable development strategy has been one of the many successful attempts to cope with the recurrent crises in a traditional manner because the challenge has presented itself in a traditional manner. The originality of the current situation lies in the fact that the political debate's focus has shifted from social to natural issues, i.e. from class relations to human/nature interactions. The traditional interpretative paradigms have clearly failed in creating the grounds for an adequate political action and in creating a link between scholarly thinking and the beliefs of the people. If in the past they have helped in constructing a dialectic and revolutionary discourse which stood as an alternative to a corporate neoliberal system, nowadays the social conflict categories – be they applied in the terms of Marxist class struggle or in the terms of inherent libertarian market social competition – are failing to satisfy a large part of public opinion in western countries.

Before approaching the socio-political specific organization problem, we need to accept that modernization and industrialization have a lot to do with a particular relation between humans and nature. The evolution of the current science-based technology, introduced in the sixteenth century, is the outcome of a philosophical negotiation with nature. Suggestively, Fritjof Capra maintains that, if Leonardo's systemic and holistic post-medieval scientific approach had prevailed on the more successful Descartes and Galileo's method, we might have experienced a different evolution of science and of the relationship between humans and nature. Capra's idea is evocative no matter how arguable his thesis is. We have mentioned it because in relation to the environmental question, we need to call for a paradigmatic shift not only in science and political analysis, but also in political action.

Thus, instead of reiterating the paradigm of nature as a political construct, it would be more useful, from a heuristic point of view, to distinguish between nature and environment and give nature a political subjective role.

Moreover, the centrality of work and labor as a transformation process, as advanced in Marxist theory, has created the grounds for treating Nature as an object exploited by the system in the same way as workers are.[31] The relation between human beings and Nature includes an issue of justice. There is already a rich literature about environmental justice, but most of it focuses on how environmental risks and harm are unequally distributed socially and geographically. Nonetheless, environmental justice may include a broader concern and apply to nature.

Arguing on Sustainable Development has been a necessary starting point for this essay on urban traffic policy. If we really want to change the way we act because we are not any longer satisfied with the urban quality of life, we cannot proceed by continuous adjustments of the current condition. Thus we need a new thinking more than a new practice. Traffic policy - indeed a "traffic revolution" - is a case study whose implication may be extended to several other issues of urban and environmental policy. It is also an attempt to link theoretical thinking to action.

[31] I do not deny that recent and less recent Marxist and Marx-inspired literature has treated this issue at large producing an extensive bibliography. However, the humanity/nature relationship has occupied neither the core of scholars' concern nor has won the limelight in the political debate.

Chapter 3
Technologies, Problems, Solutions

Look at what realists have done for us. They have led us to war and climate change, poverty on an unimaginable scale, and wholesale ecological destruction. Half of humanity goes to bed hungry because of all the realistic leaders in the world. I tell people who call me 'unrealistic' to show me what their realism has done. Realism is an outdated, overplayed and wholly exaggerated concept

Satish Kumar

Abstract Even if, year after year, traffic congestion is becoming more and more unsustainable, governments stubbornly commit the solution of the problem to the same people who have caused it. Civil engineering applied to all traffic problems has become the proxy of a missing ideology. Traffic and transportation plans are elaborated only by civil engineers who determine how urban mobility policies are to be conceived. Hence, the problem is set around solutions that are known and chosen prior to any open analysis. It is apparent that civil engineers have not been able to solve the problem in the last 40 years, but in this period they have established and reinforced powerful professional guilds which intertwine with industrial and political lobbies. The solutions they apply to traffic are likely to be the real cause of traffic problems. However, contemporary mobility policy also embraces issues such as: (a) health risks; (b) environmental responsibility for global warming; (c) social justice which regards how mobility-related risks, costs and benefits are distributed; (d) economic justice concerning the appropriation and redistribution of public monies. It also concerns an epistemic analysis regarding knowledge, science and the use of professional skills and competences. If we apply new concepts in cost-benefit analyses assessing the effectiveness of traffic plans, we would be able to achieve better results. We would understand that public and private decision-making is trapped in a vicious circle that involves an escalation of traffic, similar to the inflation process. It is exactly the opposite of what should be done, i.e. preparing a traffic

C. Poli, *Mobility and Environment: Humanists versus Engineers in Urban Policy and Professional Education*, DOI 10.1007/978-94-007-1220-1_3,
© Springer Science+Business Media B.V. 2011

reduction sequence. Traffic inflation is induced by the transportation and infrastructure industry's need to continuously grow. These considerations support two final proposals on how to approach traffic policy in an alternative way.

Keywords Technology • Civil engineering • Solutions vs. analyses • Traffic inflation • Utopia vs. best practices • Professional ethics • Professional guilds • Approaches to pollution reduction

Introduction

In the first part of this chapter I discuss some situations and express the feelings involved in traffic policies and in general in government decision-making related to high environmental impact constructions. In the second part I argue the economic evaluation applied to approve new traffic infrastructures.

Traffic Policies: Technicians and Politicians

A Traffic Jam

That evening it was raining cats and dogs. At seven – which in Italy is still rush hour – I drove my car to a dinner party to celebrate the decennial of an association. It took me more than an hour to cover ten miles between my house and the restaurant. I was living in the center of a typical middle size European city. The restaurant was located at the extreme urban periphery, in a place that until 20 years ago was itself a little town with its own economy and identity. Today, people hardly remember the name of this little town as it is lost in an anonymous suburb. In the rain, the traffic proceeded bumper to bumper. Everybody was late to the meeting. When it rains, traffic jams are customary. Nevertheless, the invitees were all surprised and the traffic jam issue took over most of the evening conversations. The invitees – mature and highly educated people, some of them holding influential positions in the society – could not get rid of a reiterated tune: road shortage. "It is a shame that in a civilized country we get into such traffic jams"; "they should build a highway joining …"; "It would be enough to construct a new intersection"; "In Germany … in France … in the US … this is not supposed to happen!" Guests, who considered themselves environmentalists, argued that a generic public transportation system – which the others were supposed to use – would have miraculously solved the problem.

A Solution Preceding the Problem

To overcome the boredom of those idle conversations, I tried to remember where most of those people lived. A large number of them lived in the city center.

Then I considered how many restaurants could have hosted that group. I counted at least seven or eight of them. I also remembered that until 10 years ago there were many more restaurants in the city center. Thus I came to three conclusions.

First, if the organizers had chosen a restaurant in the city, a large part of the invitees would have walked to the meeting or they might have taken a bus or a cab. They might have even driven their own car for a few kilometers and still avoided the chaotic suburban traffic. The lack of complaining invitees traveling on the oversized suburban roads would have made traffic lighter. Or, from another point of view, it would have not been necessary to continuously enlarge and multiply those roads. Road expansions are what governments have done – the only thing in traffic management governments have done for a half century – without achieving any success in reducing congestion.

Second, everybody saw only the problems that had traditional solutions. It would be more correct to say that they knew the solution without knowing the problems. The guests identified the problems by looking at them through the filters of the solution they already had in mind. In other words, because they all were undisputedly convinced that road construction was not only the easiest, but also the single possible solution, they just focused on how best to build the roads. They did not take into consideration, even as a hypothesis, the idea to possibly change people's mobility patterns. When I tried to introduce doubts in their belief and steer the conversation in this direction, they aggressively silenced me.[1] In the best case I was ridiculed, although with good manners that the educated upper middle-class commonly exhibits.

Engineering as a Proxy of Ideology

Eventually, everyone imagined idyllic situations realized somewhere else, in an imaginary foreign location. Professionals who deal with the urban traffic problems know that the same traffic congestion takes place almost everywhere in cities. The problem is not so much a common psychological attitude that brings us to create an ideal image. When we imagine and describe a utopian solution, although unreal, we pretend it really exists in a hypothetical "somewhere else." This mental pattern – explained by the very meaning of the word *u-topia* (no-place) – makes people confident and spurs them to proceed toward an actual solution. The drama, instead, was that the fellow-invitees did not want to escape their rigid schemes. They replicated a conceptual model that they imagined as perfectly developed "somewhere else," that was eventually identical to their own place, hence not a real utopia. In fact they showed a dramatic lack of creativity.[2]

---

[1]I must confess that I was not very polite in arguing their beliefs in that occasion. At a convivial dinner party people legitimately want to relax and repeat idly their biases and their convictions. The problem is that prejudices and lies, if too often repeated, become half-truths and an utterly shared discourse.

[2]Note that, assuming that a project really exists "somewhere else" is the opposite of a utopia in which case you assume that the imagined project exists in "no-place", thus it is absolutely original.

As a matter of fact, this attitude is exactly the opposite of a utopian thinking and rather conforms to the currently more habitual "best practices" approach. To make it simple – as we do not need to get stuck in complicated philosophical arguments[3] – we can conclude that the collection of "best practices" is anything but creative and is based on the assumption that there is nothing new on earth. Thus, we are neither allowed nor encouraged to propose doing something which has never been actually applied before somewhere. If someone, in order to call for change, needs to claim that there is a real place – so not a utopia, i.e. a no-place – where their ideas are implemented, they refuse the possibility of a creative thinking.

Back to the dinner party: to even conceive a possible change, the fellow-invitees needed to think that an efficient project existed and was actually implemented somewhere. They did not consider any possible radical change and argued as if they all were experts in traffic engineering, in bridge and road construction, in transportation means design, and of course in public administration. As a matter of fact, they all were authentic experts in transportation engineering, indeed. Civil engineering has become the only available solution, and it was deeply assimilated in everyone's mind. It stands as the proxy of an ideology. Perhaps, the invitees did not know the technical and operational details, but everyone could competently repeat concepts and propose tools that civil engineers have utilized in the course of the last century. They were absolutely fluent in civil engineering language, but could not speak any other when dealing with traffic. Moreover, they were outright convinced that there was an easy solution to the traffic congestion if only the people in the government were not obtuse. I was surprised because some of them were economists, others lawyers and there were also sociologists and philosophers. They all were highly qualified in their fields. However, they refused to even allow a possibility to apply their own discipline tools to analyze the problem. They did not yearn for the opportunity to espouse – just as in a convivial conversation – their competence to address such a complex issue as the traffic problem. Not to mention the most predictable perception of urban settlement that was interpreted as commonplace a century ago, although they were the living proof of completely changed physical, technological, and social circumstances.

[3]I will return to the best practices issue in Chaps. 4 and 6. The Utopian approach has been typically held in great contempt by materialist and Marxist philosophers who endorsed the famous Marx's 11th thesis on Feuerbach which goes: "the philosophers have only interpreted the world, in various ways; the point, however, is to change it." The Frankfurt School has tried to rehabilitate the notion of Utopia as a change tool. According to Horkheimer, Utopia has two main meanings: a critique of the current situation and a project for the future. "The materialist longing to grasp the thing aims at the opposite: it is in the absence of images that the full object could be conceived. Such absence concurs with the theological ban on images. Materialism brought that ban into secular form by not permitting Utopia to be pictured; this is the substance of its negativity. At its most materialistic, materialism comes to agree with theology" (from Adorno's *Negative Thinking*, quoted in Benzaquen 1998).

Traffic and Ethics

The urban traffic problem embraces ethical evaluations regarding: (a) health risks; (b) environmental responsibility for global warming; (c) social justice which regards how mobility-related risks, costs and benefits are distributed; (d) economic justice concerning the appropriation and redistribution of public monies. It also concerns an epistemic analysis regarding knowledge, science and the use of professional skills and competence.

In a city that is a place where "*we move around*" instead of a place where "*we live*", a change in mobility patterns is likely to alter both the property values and the quality of life for a large part of the people. Sometimes the building of a new road, or just the inversion of a one-way direction, implies heavy consequences for the lives of the citizens directly affected. An area that was once quiet can become noisy; a place that was isolated and hard to reach can benefit from better accessibility. If the access to retail facilities, dwellings and offices is modified by the construction of an infrastructure, some will win and others will lose. Contemporary cities – mobility being their basic characteristic – are more sensitive to these kinds of changes than in the past. Nonetheless, we do not pay enough attention to them. In Pareto utilitarian terms, a decision is ethically acceptable if, in consequence of it, nobody opposes (loses) and at least someone gains. Instead, these ethical redistributive problems are almost completely alien to traffic discourse. We care a lot about other social justice issues such as racial discrimination, equal opportunities for women, and so on, but do not have the intellectual tools, and even less the mental habit, to deal with this current crucial aspect of social and environmental justice.

In several cases, it happens that infrastructure construction is opposed by activist groups. This can be correctly considered a proof – though indirect – that ethical and redistribution concerns are taken into consideration in the decision-making process. However, in their resistance to projects, people focus on the technical aspects (how to reduce environmental impacts, the design, the place where it should be located, the cost and the benefit in an undefined time span, etc.) rather than on who wins and who loses and how much. The ethical and redistributive concerns are kept implicit and are confused among other issues, so that they cannot be properly examined. Moreover, environmental goods and concepts such as precautionary and intrinsic values[4] are even more rarely assessed or just taken into consideration.[5]

[4]In customary discourse, which is mostly derived from economics, we assign the economic good only the so called "use value", that means we value it according to its immediate capability to satisfy some of our needs. In environmental studies, and specifically in environmental ethics, other values are typically considered, i.e. precautionary, inherent, intrinsic and existence values. See Singer (1991) and an abundant specialized bibliography. The importance of economic values besides use values have been treated by pioneer ecological economists such as Kenneth Boulding, Nicholas Georgescu-Rougen (1971), and Herman Daly just to quote some of them.

[5]In 2000, I was in charge of a cost-risk-benefit analysis regarding the construction of a dam and a reservoir in the Alps. The project was very controversial: as expected it was opposed by environmentalists and supported by industrialists. The situation was even more complicated because part

These issues should entail consequent analyses and policies. Firstly, when mobility patterns are modified, the question of justice should immediately be posed and considered a serious concern, since the contemporary city is organized on traffic much more than it was in the past. Secondly, we should comprehend why, when a new traffic plan or new infrastructure is to be built, more and more protests are likely to happen, at least in the US and Western Europe. Although these frequent protests seem to be focused on specific technicalities of the infrastructure projected, their growing violence is justified and explained by a social justice redistributive problem. Violence is even more likely to happen because we lack a political language shared by the grassroots, that is capable of dealing with environmental and transportation issues approached as political, social and ethical ones. For as long as major social problems have been human dignity of workers, the right to be employed, economic development, etc., scholars have elaborated ideologies capable of framing the political debate also at the grassroots level.[6] On these ideologies, also the political debate was organized and, with it, the political organizations. No doubt, that this discourse is the dominant one and it is still necessary. However, we lack an environmentalist ideology and a shared political language that allows us to perceive current crucial problems, such as traffic and environmental crisis, in the most convenient form to deal with them. It is not incidental that in the most developed countries, citizens are more likely to protest than in backward ones that are still struggling with economic development. The social justice problem is even more serious because, to modify traffic patterns, it is necessary to invest huge quantities of the taxpayers' money which could be appropriated differently.

While I was editing this essay, major TV networks reported on two citizens' demonstrations which were supported by national unions and parties: one regarded three FIAT workers who had been fired by the company that accused them of felony in organizing a strike; the other concerned a student rally in favor of more funds for university and public school. These two protests immediately became national issues

---

of the valleys' population was favorable and so were some of the local governments involved. I suggested that the commissioners hire a moral philosopher to help in devising a coping-strategy and solving the decision-making problem. They (surprisingly) accepted. First, I asked the ethicist to identify the moral values involved in the decision-making. Moral values were implicit in the assessment of impacts on the natural and human environment (environmental ethics), in "who had the right to decide" (political ethics) and in cost, risk and benefits distribution (social justice). I also asked the philosopher to describe and predict the parties' conduct according to the values that they were not capable of expressing rationally but which were heavily affecting their behaviors. In this way I could understand the consistency of their behavior and it became predictable. In raising the question to the applied ethicist and in locating his competence in the planning team, I was inspired by Shrader-Frechette (1991). Then, I asked the moral philosopher to help create a shared language so as to ease the dialogue. The experiment was successful, although I realized that: (a) we still lack applied ethicists who are properly trained to deal with activists, professionals and decision-makers; (b) activists, professionals and decision-makers are not used to speaking the language of moral philosophy because their beliefs and languages are heavily value-loaded, but they do not realize it and prefer to use their familiar technology notions.

[6]See the rich literature about advocacy planning and citizen participation. Between the 1960s and the 1980s, this literature was mainly focused on social justice, then the environmental issues have come into the limelight.

because they evoked the century-old struggle for more worker and student rights. The local events were part of a national and international ideological discourse that everyone could understand, be part of, and easily join a side. This fighting was also quite safe since unionists and students were protected by well-established organizations who could stand up for them in the media and possibly in the courts. The protesters were not saying anything new. Whoever was watching the news already had their own opinion that would hardly change because of the protest or the way the news was reported. The two protests were unremarkable, but it seemed normal and easy to speak about the two rallies. On the contrary the real news, almost completely neglected by national media, would have been the many protests that are taking place all over Europe and North America to oppose hundreds of high environmental impact projects. Most of these projects concern mobility infrastructure constructions such as roads, railways, airports, parking lots and shopping centers, not to speak of waste disposal (which also includes road traffic as it requires moving thousands of trucks) and all kinds of hazardous plants. The lack of a language and of an ideology, capable of synthesizing them in a unified thought and movement, hampers them from becoming a national issue. Nonetheless, if we calculate how many people are currently involved in environmental advocacy, we would learn that they are much more than the militants in unions and traditional parties, and – for the most part – are much more active. Unfortunately they do not have a voice because they still lack a universally comprehensible language.

Thus, if nothing else, these considerations explain why traffic policies raise so much social conflict and why a traffic environmental revolution may become the spark for a more general social and political change.

Solutions as Causes of the Problem[7]

"For how long have we been defeating Germans!" This exclamation of Tacitus can opportunely be applied to urban traffic plans. Ironically, the Roman historian ridicules the triumphs the Romans had been celebrating for two centuries for their numerous victories over the Germans who, nonetheless, were still there threatening the *limes* (frontier). Those premature triumphs resemble the ceremonial and cheering inaugurations of new highways, urban rails, bridges and intersections, beltways and overpasses. As in Roman history, we might raise the question: were the Germans such a barbaric people that the Romans could not tame them? Or were the Romans such a decaying power that they could not colonize any other country? This essay

[7]The specialist reader can easily conjecture how this part is directly inspired by Michel Crozier's research on bureaucracies and system behavior (Crozier and Friedberg 1980). "Forget about solutions" and "Learn thinking differently" are Crozier's mottos meant to change inefficient bureaucracies and social systems. It is easy to recall also how traffic was one of Herbert Simon's main sources of inspiration, although his "solution" goes in a different direction: i.e. system analysis rather than the socio-political approach. See also Crozier (1995).

does not deal with Roman history, but the question we have raised is similar. Is the urban traffic problem just unsolvable or, more realistically, we cannot face it because the tools we use are old and inadequate? Might those tools be the very cause of the problem rather than the solution? Is it possible that we have unrealistic expectations about rapid access to places, the infinite availability of land, and in general, on further progress in a direction we chose a few centuries ago? Almost 40 years ago we sent a few men to the Moon. Recently we have sent spacecrafts to Jupiter and Mars. We have even artificially created life … and we still live in traffic jams![8]

At least 40 years of social research and polls find that citizens consider traffic as one of the most serious and urgent urban problems. For 20 years traffic has been associated with environmental deterioration and health risks. Nonetheless, year after year, it has been discovered that conditions have deteriorated in all cities. We commonly adopt the smokestack strategy, i.e. the roughest anti-pollution system by which damage is spread out to wider areas so that a presumed immediate benefit can be gained. The result is twice as negative: on one hand we extend the damage and prepare a vicious circle of traffic inflation that I'll describe later in this chapter; on the other, we overlook the possibility of more efficient environmental policies.

Four Approaches to Pollution Reduction

The smokestack approach to local environmental deterioration is the simplest and most basic level of intervention. A higher level of environmental policy would imply the adoption of depuration and disposal techniques that are almost inapplicable in the case of traffic, but are largely used in other fields of environmental policy. The third level would be the implementation of new technologies that produce less pollution, like the adoption of the catalytic converter and the use of clean combustibles. The most effective environmental policy would be the removal of the very causes of pollution, i.e. a significant mobility reduction or the development of zero-emission engines. However, possible new zero-emission technologies would not have any real effect on other heavy impacts such as road construction, car production, and the movement and transformation of oil, iron and chemicals, not to speak of the social consequences such as urban sprawl, excessive mobility, weakened relations between people and places.

When air pollution levels became unsustainable and illegal, all governments, both in the US and in Europe, have generally enforced temporary traffic limitations while continuing to trust only technological progress. As a matter of fact, the European Union enforces a law that imposes traffic limitations in all areas if the current air pollution exceeds certain parameters. Governments that do not comply with the parameters must pay a fine to the European Union. Moreover, public officers may be sued if they do not take action against pollution. Notwithstanding this severe legislation, very little has been done to contrast traffic-generated air pollution in terms of the long run and structural policies.

[8]Arguments about the Moon and the ghetto paradox was common in political science during the 1970s. See Nelson (1977).

Problems and Professionals

Even if, year after year, the traffic problems are becoming more and more unsustainable, governments stubbornly continue to act in the same way. Absurdly, they commit the solution of the problem to the same people who have caused it. They show the same frivolity of a shepherd who would entrust the custody of his sheep to wolves. A few years ago, I was directly involved in a decision-making process for the approval of a mobility plan in a European middle-size city. I found out later, doing *ad hoc* research and consulting as a professional for other diverse city governments, that what happened there is exactly what happens almost everywhere.[9] The Municipality government assigned the design of a new traffic and mobility plan to a large engineering consulting firm that was actually a branch of a multinational company specialized in transportation production means, i.e. light rails. The plan was realized by civil engineers who possessed an expertise in construction. These professionals – trained by and connected to the multinational company – already *knew* the solution to the traffic problem. We do not need to give a clue to help the reader guess what solution was identified!

It would be misleading to think that all this happened because of the unethical behavior of the professionals hired. It is not even necessary to think that the firm's executives were somehow corrupted, nor that there were politicians who might have possessed some underhanded business interests. Perhaps the only ones who could really have a clear vision of the whole process were the Mayor and the consulting firm's CEO. However, I can guarantee that both the Mayor and the CEO were reasonably honest, skilled and trained men.[10] They just lacked an educational background that would have allowed them to propose any different solutions to the traffic problem. Consequently, they could not help but trust what had already been done in the past and elsewhere around the world. This was the safest strategy for them, regardless if it was clearly driving the city to its doom.

Big Companies, Huge Solutions, Zero Intelligence

To a large extent, the mobility planning process was just a series of inevitable decisions that, once the process was initiated, proceeded in an undisputed and uncontrolled direction that was fatally predefined. No surprise then that all that really came out of the plan was a light rail recommended to be built in conjunction with a list of new roads, parking garages, and new highway connections. It was a shopping list of new constructions without any priority ranking. Could one expect

[9] See, e.g., Richmond (2006), who describes the case of the light-rail construction in Los Angeles, proves with an abundant argumentation and exhaustive documentation how useless "solutions" can be realized just because they are well politically marketed.

[10] Below, I will argue that their actions were morally reprehensible although they were not corrupted according to the common understanding of the word.

anything different? Ten years later, a new administration, which had overthrown the old one, commissioned a new plan from another company. The newly hired company was an identical twin of the other, although it was a competitor of the former multinational. What about the proposals? Indistinguishable! Except for the brand logo on the light rail cars. And what about citizen participation? Because of the absence of any meaningful choice, politicians – purposely or not – allowed citizens to dispute only irrelevant technical details that were the proxy of a consistent plan.

Public opinion manipulation was not apparent and definitely not proven. Spin doctors might have intervened to vilify in the media an alternative project proposed by activists and enthusiastically consented to by most citizens. However, I am inclined to think that the vilification process was not deliberately organized. Rather, the lack of trust in any alternative traffic policies is the consequence of a mood that has been created in the course of the years.[11]

Prospecting a solution to a traffic problem based on a single decision – such as the construction of a bridge or an urban rail – is easy and it is effortlessly comprehensible by laymen citizens. Nonetheless, most of the time, it is ineffective. A comprehensive traffic plan, based on a multiplicity of decisions, minor tools, and rationalization policies is both difficult to design and to explain. But it is the only possibility we have if we really want to tackle the problem effectively. Instead, unfortunately, we go on dreaming of the single cathartic solution. This consideration applies to urban planning and urban architecture as well. Developing countries' capital cities host a number of architects' presumed masterpieces. The fantastic bridges and the daring skyscrapers are located in the middle of nowhere, where indeed a web of human relations has been absolutely overlooked by those lazy architects and their commissioners. The Shanghai MAGLEV, the celebrated world's fastest train, is the emblem of this mindless policy: it goes from the airport to … nowhere twenty miles away … but very fast (max speed is 311 mph). The German construction firm admits that it is only an experiment, no matter how expensive and useless. However also in the western world, albeit we did not reach such an excess, several high speed train services have been questioned and opposed because they were considered ineffective and even detrimental for the economy of the towns and cities by-passed (and so cancelled) by the fast rail.[12]

What the Words Mean

Before continuing the description of how decisions are made, let us focus on commonly used terms, the original meaning of which has been partly lost. We often use words like "traffic", "transport", "congestion" and "communication" as synonyms or without paying attention to their real meaning. This happens because,

[11] This consideration goes along with what Swyngedouw (2007) claims about the victory of the neo- capitalist order and was discussed in Chap. 2.

[12] I have treated this issue extensively in a recent essay published in Italian (Poli 2009).

being focused on the known solution, we do not care about the definition of the problem, which would imply a precise use of the terms. More precise consideration about the meaning of the terms used helps in defining "who should do what" in traffic and transportation policies.

"*Tra*ffic" and "*trans*port" have the same Latin root that in English would translate into "through" associated with the idea of going/movement. The first term means "go through" and should not imply any value judgment. A traffic policy or a traffic plan is simply intended to make movement easy. Traffic is generated by the necessity, the desire, or the interest of people to move goods and/or to have access to them. Thus, the movement of people and goods requires "transport" that is composed by two words: "trans" and "port", the latter meaning "carry". Transport means "shifting things and persons from a place to another". A transportation plan or strategy should consequently be different from a traffic plan. The latter is a technical exercise that does not imply any real critique about the movements. But a transportation plan should require some economic and social reasoning about the efficiency of the movements and the opportunity of them in relation to issues such as social justice, development models and quality of life. Equally, it should include economic efficiency.

To tell the truth, the word "traffic" is now commonly used as a synonym of congestion. This is quite descriptive of our unconscious convictions: it describes our real despair about making movements easy and pleasant. "Congestion" also derives from Latin. It is composed by two words: "*cum*" that means "with/together" and "*gerere*" that means "drive/manage/run". According to this etymology, "congestion" happens when all things go together and likely get confused, fused together, indistinct. Traffic is then *congested* when everything gets messed up and we can no longer comprehend why we move and transport goods around.

The word "transport" has historically been associated with "communication," that comes from Latin "*cum*" and "*unum*" that clearly means "make one," "put together." Indeed, before the invention of *tele*communications, we needed physical transportation if we wanted to get together. Nowadays communication and transport are, to a large extent, independent from each other. Of course, our psyche has not yet assimilated the separation between transport and communication. Even less so in most traffic and transportation plans.

The Cul de Sac of Civil Engineering

After a careful review of the professional background of those holding leading positions in departments of traffic and transportation, both in Europe and the US, I found that most of them have degrees in civil and/or transportation engineering.[13] In the US very few of them have a background in planning. This preference proves

[13] In academic and professional organizations, transportation engineering is traditionally a branch of civil engineering.

that mobility organization is above all a matter of physical movements. If we want to stay in the engineers' professional guild, we might think about hiring management engineers, logistic engineers, telecommunication engineers, organization engineers and more. A shift in the qualifications we demand of transportation professionals would be a wise decision in relation to the competence offered on the job market. Even though I would still advocate for a more radical change in the competence required for the professionals involved, I would admit that we have moved on from the nineteenth to the twentieth century! However, I dare to go even further, and I maintain that we cannot design mobility plans by trusting leadership to people whose professional training is primarily technical.

For too many years the effect of this passive approach has resulted in trivially perceiving the question of mobility simply as filling buses, cars, and trolleys with the highest number of people. The blind goal is to let people move as fast as possible to where everyone thinks they want to go. People's real preferences are not seriously argued. In making mobility plans we do not even think about beginning the analyses by considering social behavior, new citizens' attitudes, economic situation, information technology, and communication tools such as cell phones, the Internet, environmental protection ethics, health concerns, and so forth. These are never priorities. In the best scenario, these issues enter the scene when the play is over. They are secondary in respect to the compulsory drive to build roads and physical – not even organizational – structures for mass transit to make more people mindlessly move. What a great idea of progress! These traffic specialists are proud to adopt models inspired by hydraulic engineering. In fact, in the traffic jargon, people often mention flows, discharge pipes, fluidity, bypass … as if human beings' behavior could be equated, not even to the one of a living organism who can make decisions, but rather to the one of an unanimated molecule of fluid.[14]

A New Leadership Needed

From this argument, I do not conclude that civil engineers are superfluous in implementing traffic and transportation policies. Rather, I claim we cannot trust them in the leadership and management of a critical situation which they themselves actually created. We should admit they have not been able to solve the problem in the last 40 years. In this period they have established and reinforced powerful professional guilds that intertwine with industrial and political lobbies. The problem of lobbies' influence is not an issue of possible corruption, back-kicks, and briberies that sometimes occur in a public decision-making that typically

[14] This approach has noble roots in the positivist belief that approaches the understanding of society as an object of scientific observation and consider social science as a technological product. The classic critique to this approach emerges clearly in the famous Habermas-Luhmann debate (1971).

implies the appropriation of a huge amount of money. If it were just a corruption problem, its solution would be easy, or at least well identified. It would be discussed in courts and political debates. Unfortunately, the situation is much more critical! Even if the administrators were honest, and the managers highly qualified, we dramatically lack an environmentalist culture rooted in social sciences and in professional ethics. We do not have the intellectual means – and probably not even the courage – to oppose an alternative approach to the mere technicality of civil engineering and the practice of business as usual. Moreover, most of the environmentalist advocacy groups also adopt a language focused on the technological aspects of traffic instead of considering the mobility problem as a social one. A social science approach and a humanistic language would dramatically help in promoting change.

The Academic Responsibilities: "Who Does What"?

Organization sociologists, experts in management and logistics, social scientists, philosophers in professional and applied ethics, economists and scholars in public finance, epistemologists and urban planners, and eventually all highly educated people, bear heavy responsibilities in keeping this piecemeal and obsolete approach alive. They are very reluctant to invade someone else's territory so that they do not even try to break the monopoly that civil engineers have established in dealing with mobility issues. In fact, the risk is to be ridiculed. Because the traffic situation is always perceived as anxiously urgent, it seems that decision-makers need to find solutions overnight. Thus, they feel as if they do not have time to consider some basic problems: Who can help? Who should do what? How can we train new experts? It seems that time is lacking to focus at least: (a) on the decision-making processes; (b) on middle and long term planning; (c) on arguing about the qualifications and training of the people in charge of urban mobility. We must admit, instead, that as time goes by, some tools become useless and even harmful. The mobility-related disciplinary paradigms of civil engineering, and techniques developed a century ago, cannot solve the traffic problem anymore. Probably they make it worse. The solutions adopted in the context of the traditional approach tend to reiterate solutions that have proved ineffective for decades. This problem not only concerns mobility and traffic. It applies to other problems, professions, and skills.

A famous Mark Twain aphorism describes this situation thusly: "if one's only tool is a hammer, then all problems look like nails." But we are in an even more desperate situation: all those who have tweezers, screw drivers, drills, organization models, information technology, land-use regulation, fiscal laws, and more, do not dare to treat the traffic problem with their own tools; instead, they expect a solution from the one hammer of civil engineering. If we try to hammer a screw into a wall the effects are devastating to the wall. But there are also profound effects on the psychological balance of the hammerer, who is afflicted with a profound feeling of frustration. What surprises me – even more than moral law and the starry sky – is the complacency of the civil engineers *vis à vis* all their recent failures! There is a good reason, though, and it will become apparent in the following paragraphs.

The Solution Is Not to Solve: Why Are Cheap Solutions Not Popular?

In traffic policy, we should intervene by opening up various options and by using a number of different tools. There is not an ultimate tool or option that could bring us to a definitive solution. However, there are many tools that could give partial results that, as a whole, would be either effective or disastrous. We do not operate this way because we do not want to. Rather we discard many possible options because it is hard to apply a method that is extraneous to the current cultural milieu of those who are officially in charge of finding a solution. Several partial policies for traffic and mobility that are apparently limited to addressing only some aspects of the general problem – such as priority given to bicycles and/or pedestrians, parking, telecommuting, car-pool/car-sharing, economic incentives, and so on – are more effective than a specific infrastructural construction intended to be the final solution. Heavy constructions do not change anything, but instead, add new problems. The crucial flaw of implementing numerous and diverse policies is that they are difficult to manage and do not require concentrated investments. This makes them uninteresting for many decision-makers who are fascinated by big business and huge constructions.

If these policies are cheap compared to highway and mass transit constructions, why do the administrators not approve at least some of them to make citizens happy and to resolve problems simultaneously? For example, the administrators could promise significant tax cuts since they would not be investing money in expensive and harmful constructions. Not even this seems possible at present. First, many citizens distrust the effectiveness of new policies in fulfilling their mobility needs, which they consider immutable, since they are not well informed on possible alternatives. Thus, they trust the traditional heavy infrastructural policies better. Second, these policies, though minimal, are politically dangerous because they might begin a process of altering the status quo. Several times I have witnessed that, when administrators tried to adopt some innovative traffic policy, others opposed them because of their dependence on lobbies and guild support. The construction business and engineers' guilds used their expertise and influence to boycott ferociously these policies both through derision and by withdrawing any political and economic support. The typical argument used by the boycotters of these alternative policies is that they are theoretical and unrealizable: "we need time and investments for them and all might result in a failure", they utter promptly. Unfortunately, they do not admit the actual failure of what has been so "practically" completed for many decades and they continue to do this without expressing the same doubts. How much time and money was necessary to build beltways and light rails? Many companies, professionals, unionists, and politicians take advantage of pollution, traffic, and of absurd mobility patterns. What is even worse is that these leaders make their living by speculating, not only on traffic, but also on traffic inflation, i.e. on a system that determines a constant increase of traffic.

Economic Evaluations

Traffic Inflation and a Traffic Reduction Sequence

Historically, public and private decision-making is trapped in a vicious circle that involves an escalation of traffic, similar to the inflation process. It is exactly the opposite of what should be done, i.e. preparing a traffic reduction sequence. One could argue that this is the typical unlimited growth sequence that the capitalist system requires and applies to traffic and transportation structures as well as to any other industry. This argument is only partly acceptable. In fact, capitalism has proved very flexible when necessary and, in the past, has been able to totally transform – or even get rid of – some industry. Thus, we may find a "solution" to traffic problems even inside the capitalist system. This theoretically possible self-reforming of traffic policies is altogether unlikely since transportation and traffic are one of the main features of industry and of people's life-style. To change them, we need a thorough transformation of many mental schemes that most of us do not yet question.

Traffic inflation is induced by the necessity of transportation and infrastructure industry to continuously grow. In political economics we have studied that a slight (creeping) inflation – between 2% and 3% per year – has positive effects on the economy because it may stimulate investments, credit and production. In economics, inflation – which means filling something with air, i.e. metaphorically with nothing – occurs when supply fails to meet increasing demand. Scarcity necessitates rising prices. Sometimes, as prices grow, unaccountable authorities may react by increasing the quantity of money in circulation, while supply (production) stagnates or grows at a lower rate than demand. The consequence is a further increase in the price growth rate. Applied to mobility, we can conclude that, if we increase the supply of new traffic infrastructures, and fail to bridle the additional demand they generate, new mobility infrastructures will soon prove necessary. Hence, when satisfied administrators delightfully inaugurate a new road (or a bridge or beltway) which temporarily removes a traffic congestion problem, they should simultaneously explain to their constituency what measures they have adopted to counteract the extra traffic that additional infrastructure is going to create. Only then can the construction of a new infrastructure be justified.

This approach is difficult to explain to a lay audience. In fact, in the last two centuries, we took for granted that an increase in the mobility of people and goods would unquestionably create positive effects on cultural and economic progress. Nowadays, we can safely believe that a further increase in the overall physical mobility – and consequently the infrastructures that make it possible – does not add any extra value to society welfare. If we calculate all the values involved, not only the monetary ones, every physical mobility growth – both of people and goods – generates a loss rather than a gain. From a mobility growth, taken as an end *per se*, we should move on to a selected growth of effective mobility, which might suggest a reduction of mobility in absolute terms. If someone wants to disagree with this statement, they are welcome to bring scientific evidence. Even though some research has been carried out on this topic, it is marginal in the current discipline debate. Whoever protests against a further growth of mobility and mobility inflation has to fight entrenched and unquestioned biases.

Costs vs. Benefits, Discomfort vs. Quality of Life

Even adopting the typical utilitarian cost/benefit analysis, we can prove that most of the infrastructures are neither cost efficient nor cost effective. When we calculate the total costs of a mobility infrastructure, all too often we overlook or undervalue environmental costs.[15] For example, we do not consider the damage caused to Nature by careless exploitation of quarries and rivers to produce construction materials. Because these are economic activities and produce an income, they are calculated as benefits rather than costs. Moreover, we do not calculate the reduction in property values of units located near the new roads nor do we include the costs of distress on people caused by never ending works. We might justify the cost of distress assuming it is lower when compared to the expected benefit. But this may be true only in the case that we refer to the single infrastructure. Most of the time, construction sites are managed without consideration of the distress they cause to the population, because the sites are considered merely a temporary discomfort which will be compensated and justified by an unquestionable blessing to come. Today, with commonly adopted technologies, the construction phase itself produces a high environmental impact that is generally underestimated if not completely neglected. To asphalt 300 ft of a road in my neighborhood I have counted the deployment of

[15] In the 1970s and the 1980s environmental studies were innovative as previously described in Chap. 2. At that time environment economists made a consistent research effort to re-adapt the political economy scientific approach to the new situation. They elaborated sophisticated evaluation and assessment methods which included environmental values. Also statisticians cooperated in this effort to calculate more precisely the welfare and the real value added created by human action. It was proposed that companies and governments approve environmental budgets along with financial ones. Unfortunately, this generous intellectual effort did not produce widespread applications and the models elaborated have not been introduced neither to support decision-making nor as a political strategy to prove the presumed overall negative impacts of the new constructions. This valuable research was not allowed to leave academies, thus it was never commonly applied and even less created a political discourse. Eventually it lost popularity also in academies.

three trucks, a caterpillar, three pressing machines, two on site asphalt mobile processing trucks and God knows what else. All this without any concern for the noise and the toxic substances produced! Construction companies are rarely requested to adopt low environmental impact working procedures.[16] If we sum up all the distress caused by hundreds of construction sites simultaneously open, the cost/benefit analyses should take into consideration the cumulative costs. The total distress is likely to be superior to the sum of all single distresses usually calculated by cost/benefit analyses.

If we are uncertain about the effectiveness of a new infrastructure whose benefit will be enjoyed in the future, we should calculate the time discount rate in relation to the age of the population besides the other usually considered issues. Even if the overall expected benefits are superior to costs, we wonder if the costs are legitimate. In a condition of uncertainty about the real opportunity of the infrastructure, the oldest people, who are not likely to enjoy the benefits for a period long enough to pay off the distress, should not be forced to accept the decision. Since from a utilitarian perspective it is predictable that some will gain and others will lose, all the effects should be considered, and the decision should not be made if someone suffers. In simpler terms, how happy can my 96 year old grandfather be about living the next 5 years inundated by noise and dust from the construction site of a road that will free the village from crossing traffic in the next 10 years? The single case can be sad, but irrelevant. Nonetheless, the distress caused by hundreds of work sites should be calculated for all the population involved. So, the question should be raised this way: if I am going to spend a quarter of my life in the middle of road construction sites in the hope that in the future my family and I will live better, how much do I value my uncertain future happiness in relation to present distress? The answer would be favorable to the road construction only if the future benefits will be certain and high, and mainly if a large part of the population is convinced that the expectations about the benefits will be fulfilled and those benefits are indispensable for a better living.

Stop the Growth: Virtuous Circles

In economic policies, most deflation measures are enacted to invert trends, rather than to apply authoritarian price controls. Similarly, we should prepare virtuous sequences also in traffic policies. A political campaign aiming to shift resources from construction and transportation to other sectors, would find allies to slow

[16] As a matter of fact, in most affluent countries, civil engineers have been asked to develop and apply techniques to reduce impact and people's discomfort during the construction phases. However, in very few cases and only after very strong protests, has an infrastructure project been rejected for the reason that it would have bothered residents during the construction period. The bias that any new construction is good *per se* and adds to the general welfare for the time coming is not likely to be questioned.

down the crazy run of mobility growth. From an environmental point of view, the question is posed in a similar way. An initial reduction of 5–10% of traffic would make it easier to operate public transportation, or an alternative low impact transportation method. Mainly, it would prepare virtuous sequences that would progressively ease the situation. It would not even be necessary to reduce the traffic quantity. A good result would be to maintain the overall mobility at present levels instead of favoring an increase of movements created by the same "solutions" that are engineered to eliminate congestion.

If the Mayor of a suburb, thanks to a central government funding, builds a new road or a rail that allow the inhabitants to reach downtown jobs and services more easily, he provokes two unwanted consequences. Neither he nor the citizens are likely to realize what it is really happening. First, the improved accessibility makes it more comfortable – *coeteris paribus* – to live farther from the major jobs and services locations, e.g. the downtown. As a consequence, we have a growth in the values of the estate and of the developing lands located farther from the downtown. The government money seems to have brought a net gain to the community. Instead, this fake advantage is paid for in different terms: First, there is an overall increase of mobility demand as we make it convenient for people to live near the places favored by the new infrastructure. Then, because more people will move in these places, soon we will need more resources and more infrastructure to satisfy the mobility demand induced by the new road or rail. The final balance is dramatically negative as proven by the last 50 years of urban history.

Two Operative Proposals

We should enact a law that would compel every administration approving a new traffic physical infrastructure to prove that it has also enacted counter-balancing measures to reduce the mobility induced by that infrastructure.

A few years ago, while I was involved in political decision-making, I suggested two actions regarding new mobility plans. First, an administration should design and adopt a zero-growth traffic plan. However, notwithstanding the adoption in principle of the zero-growth option, it might be necessary to approve a new infrastructure that will increase the mobility supply as a response to some mobility needs determined by decisions previously made. When this happens, we need a law that allows governments to approve infrastructure only if they prove that they have also enacted measures of traffic containment able to offset the potential extra traffic generated by the new infrastructure. This is based on the rationale that new infrastructure not only matches the present demand, but it also generates an increase in mobility demand. This vicious cycle should be broken. Thus, if we build a new road that permits 10,000 more cars to drive daily, we must also ensure that proper consideration has been given to the possibility that the new road will increase mobility demand of 5,000 more. Second, an act would require that at least 1% of the public budget invested in mobility physical infrastructures that increases the mobility supply, is devoted to activities and projects focused on reducing the mobility demand. To achieve this goal and give the promoters some idea about how to act, I invited specialists to propose actions for effective mobility growth reduction.

I did not expect an immediate and effective application of this new traffic reduction policy. Hence, I was not too disappointed that none of my proposals were implemented. I only hoped to present and begin a new way of thinking about traffic. Not only has this not happened, but both politicians and activists preferred to go on proposing a series of piecemeal "solutions" to the general problem disregarding a possible more comprehensive approach. We cannot instantly resolve all transportation, congestion and pollution problems generated by an inefficient traffic policy. Our immediate goal is just to reverse the current trend. Reducing the mobility demand is already a viable solution because it prepares a virtuous process.

Conclusion: The Problem Survives the Solution

We cope with traffic congestion problems in a very violent and totalitarian manner. To manage traffic effectively, instead, requires soft actions that nonetheless are really revolutionary. In a metaphor, if we liken traffic to a minor sore throat, then we act by immediately removing the tonsils and proceeding with a more invasive surgery. Nonetheless, we know that most of the time it is better to let a sore throat heal on its own, even without the use of antibiotics.

In most traffic and mobility policies, professionals still act as if each part of the plan is designed with little or no regard to the other sections. Even in those cases where planners take into consideration other aspects of mobility, most of the time, money and concern about road and infrastructure construction overwhelm any other matter. In Chap. 6 I will comment on the Northern Virginia Transportation Authority Plan in which not $1 has been directly appropriated for bike routes, notwithstanding they have been mentioned several times in the document approved.

Neither the side effects of the decisions made, nor the consequences of the decisions in a broader perspective are seriously taken into consideration. For example, when traffic engineers design a new road, they do not care about the consequences of new transportation structures on the urban form and on communities. Most transportation planners assume all these induced problems as something to be dealt with later by other professionals. Instead, these induced problems are an integral part of the mobility problem. This approach seems to be efficient and practical because it reduces the general problem to a series of little projects, each of which can be perfectly designed. If the outcome is that all together these perfect projects generate a disaster, nobody is responsible. The decision-makers and the designers keep themselves convinced they made the perfect thing so that they can blame all the others for being incapable and foolish.

Chapter 4
Mobility and the Corporatist Society

Our knowledge, as well as our ignorance, at any time and on every issue, tends to be opportunistically conditioned, and thus brought to deviate from full truth. In every epoch and every problem, this opportunistic tendency operates also in our scientific work, if not critically scrutinized.

Gunnar Myrdal

Abstract Social change is a difficult task. One of the barriers to change is the resilience of the professional corporations that have organized to solve problems. Habitually, after having been operating for some time, an organization (in our case the community of engineering companies) no longer pursues its original goal. The new real goal becomes to keep the organization existing, working and possibly growing. It means that the organization is pursuing something it has not been designed and established for. Because the organization mainly strives to outlive and grow, its goals and the means to pursue them shift from being merely technical to being substantially political. However, some new problems outside the traditional competence of the technician occur and need to be tackled. The technical corporations react in two ways: (a) to subdivide internally; (b) to turn to interdisciplinary collaborations. This argument leads to a crucial question raised in this essay: should we move from a traffic policy occupied by technicians to a political environmentalism which opens to a discussion on urban settlements? The problem includes ethical considerations about both the single engineers' activity and their corporation policy. Environmentalists are responsible for not questioning the corporatist approach with enough conviction. But the ethical issue also concerns the personal goals of the engineers (and technicians in general) who act without taking into consideration the general consequences of their action and disregard the possible predominant negative side-effects of the technologies they employ.

Keywords Social change • Professional guilds • Technology and politics • Professional ethics • Multidisciplinarity • Epistemology • Politics of science • Ethics • Holocaust • Banality of evil

C. Poli, *Mobility and Environment: Humanists versus Engineers in Urban Policy and Professional Education*, DOI 10.1007/978-94-007-1220-1_4,
© Springer Science+Business Media B.V. 2011

The Sociology of Organization and Traffic Policy

A Definition of the Problems

When attempting to confront and resolve the more dramatic problems of the city, we can no longer refer to and rely on old definitions. First and foremost, it is the difference between cities and countryside areas that formed the foundation of sociology and traditional urban studies. The torn Toennies' *Gemeinschaft vs. Gesellschaft*-paradigm, on which generations of urban sociologists have hinged their basic education, is now misleading. If we still want to use it somehow, we should at least deprive it of any geographical content, namely we should forget about identifying the *Gemeinschaft* (community) with the rural countryside and the *Gesellschaft* with the city. Even the idea that the city is a place of residence, separate from the location of production, is a model ready for the trash basket. The perception of various parts of the city – the historical center, the various neighborhoods – has dramatically changed. Consequently so too has the idea of "citizenship" – that is, the right to participate in city life. City residents have become a fraction inferior to the numbers of visitors and tourists. Moreover the same city dwellers behave like tourists and visitors because of their continuing visiting of places that they do not belong to.[1]

In spite of this dramatic change occurring in the last 50 years in the relation between city dwellers and territory, most of the official data analysis, research and mapmaking – including the technologically sophisticated system elaborated by the Geographic Information Systems (GIS) – that are now customarily employed to cope with city and traffic planning, lag seriously behind. I am perfectly aware that advanced research, both in political geography and in data analysis, is aware of the changes that intervened in urban mobility patterns. Some scholars have forecast new scenarios based on an even more widespread mass adoption of telecommunications that would reduce the need for physical mobility. However, when decisions are to be made, the policy-makers act like the drunkard, who stops under the streetlight to look for his lost wallet, not because he thinks that he lost it there, but because there is more light which makes the search easier. Instead of discussing and searching what is relevant now, those researchers look where it is easier to find conventional information no matter how irrelevant the old data are, how ancient the analyses, how obsolete the technologies, how ineffective the interpretative paradigms. Unfortunately, we are not simply talking about friendly drunks. This approach is sustained by influential industrial and professional lobbies whose goals encourage

[1]Zygmunt Baumann (1993), describing tourist behavior in cities, highlighted their lack of belonging to any place. Also nomadic suburbanites, driving around all day through looking for food and services, do not recognize any place as their own.

passivity. Just bringing up this problem would be revolutionary, since it goes against numerous establishments that have no interest in changing perceptions. These interest leaders, in one way or another, benefit from their position, no matter how trivial, and inhibit change in the perceptions of cities and of the world. Also intellectuals bear a responsibility when they are incapable of exploring unknown frontiers, of elaborating new ideas, fearful of being faced with the risk of making errors or being criticized. The responsibility, naturally, also falls on the political system that cannot manage to produce personalities or contexts capable of breaking the vicious circle of obsolete skills.

The Time Lag in Organization Change: The Case of the Web

The problem concerning the gap occurring between intellectual elaboration and everyday discourse, the one that is shared by the largest part of both citizens and decision-makers, is twofold. On one hand, we can easily and ungenerously blame intellectuals for their incapability to make themselves understood by the public. We may go even further criticizing the scant interest they have in acting politically and in messing with other actors in the political arena. But this is just part of the problem of the time lag in introducing new concepts into general discourse. In fact, on the other hand, social change is typically slow to occur, much slower than the intuition of new ideas or the application of new technologies. "Rationality works much slower than most people think and even then, fallibly" as Lakatos (quoted in Currie and Worrall 1980) puts it. In the eighties, scholars were already speaking about the *Informational City*, from the title of a well-known book by Manuel Castells (1989)[2] and the idea that telecommunications would modify city form and lifestyles has been circulating for at least a quarter of a century. We should not be surprised, though, if, in reality, little has changed until now in the way we face urban problems and mainly traffic and mobility policies. Social and mentality change requires a lot of time. We think that the Internet has revolutionized our habits and behaviors. It is definitely true from a technical (and superficial) point of view, as we all use it customarily. However, we must admit that it has not yet affected most of our inner attitudes and unconscious beliefs. It will take another generation before people can fully assimilate Internet potentialities. The process is made even slower because, besides improving people mentality, it will be necessary to change also the organization and physical structures that affect everyday life. Physical and social inertias work together making change even slower. To exploit and adopt all the opportunities offered by the web, it will be necessary to reform, or better revolutionize, the

[2]Castells returned to the topic with several publications in the course of the 1990s (2000a, b, 2004).

feelings of most people and all their mass organizations, such as public bureaucracies, companies, schools, hospitals, probably even religious congregations, and so on. Last, but not least, mobility patterns, and the consequent city form, will also be modified by world-web technology in the course of the time.

This revolution can either be just a technical one that does not necessarily involve a profound change in social and ethical values, or it may trigger (or contribute to) a possible ideological revolution, especially if associated with other circumstances such as the environmental crisis. It is possible, in fact, that it might cooperate to accelerate change occurring in different contexts, which originates e.g. in the relations between nature and humans, in the more traditional class struggle, or in individual freedom rights. A technical revolution, such as the widespread use of the web, is typically a shallow one: it is circumscribed in the technical domain. An example of a technical revolution, free from ideological implications, was the spread of mass motorization. Even though the car technology has dramatically changed our cities physically and has affected our daily behaviors, it has not had direct and palpable effects on the social structure and on classic political issues such as equality, employment and labor rights, individual liberty, and so on. Very few, mostly marginalized, authors have studied urbanization and mass motorization from this point of view, nor has this issue ever been picked up as a crucial one in political competition. This is quite understandable because, at the time mass motorization began and eventually imposed itself as the basic organization structure of our cities, political attention was focused on development issues and on their implications for social justice and income redistribution. The situation is different today and the technological change fostered by web navigation might have political consequences that have not been taken into consideration regarding mass motorization. However, we should note that it took a long time to get used to mass motorization and the consequent mobility patterns. To a certain extent, European cities are not yet completely adapted to a car-oriented settlement and many people still refuse the idea of an American-like urban settlement based on the idea of mass motorization (Ray and Anderson 2000). Notwithstanding the slowness of change, the Internet has already shown the way and we need to think of our cities as a product of this technique rather than as an outcome of the obsolete mass motorization model. Therefore, the current intellectual challenge is to imagine the Utopia of the Internet city that will be available to the next generation in all its potentiality. It does not matter at all if we modify this foresight in the course of the years. This imagination will help us to begin a traffic revolution we need right now, and it will give us strength to oppose the old schemes of infrastructure building that are destroying the environment. Old-fashioned physical traffic facilities construction is making people's life uncomfortable in the present, at least to the same degree as a new traffic policy approach will cooperate to improve most of people lives in the future.

Change is a difficult task. If we merely call for change, it does not help in beginning a process of change. Hence, I will try to give some clues on how to prime a change process in the following chapters. In the next paragraphs of this chapter, however, I will discuss how urban and traffic problems are identified and formulated beginning from basic considerations drawn from epistemology and sociology of organization.

Corporatist Society and the Choice of Urban Problems

Incompetence is not always utterly negative. As Paul Feyerabend brilliantly pointed out: "non-experts often know more than experts *and should therefore be consulted*"[3]. "... prophets of truth (...) more often than not are carried along by a vision that clashes with the very events the vision is supposed to be exploring" (Feyerabend 1993: xiii). Incompetence is actually necessary to be open to new visions. Currently, the problem is rather the excessive competence in traffic planning, coupled ruinously with the inability to collate the various problems. As for mobility, research is completed concerning the flow of traffic, and calculations are made that estimate how many people go to one specific place or another. However, the problem remains that, most of the time, these data are neither interpreted nor examined in the context of the dwelling population's characteristics. Urban planning and mobility each take their own separate paths without ever meeting or communicating.[4]

In public administration – and to a lesser extent also in the private sector – each problem is resolved by an *ad hoc* professional corporation. Thus, over time, various professional guilds have formed to keep control over the problems. Guilds have elaborated well-defined codes of behavior and have eventually evolved into organizations of self-protection and self-promotion. The process originates when a professional-scientific-technical working-group is established to deal with an external challenge. Subsequently, the group consolidates in an institutionalized structure and it takes possession of the problem. The group that has transformed into a corporation or a guild, continuously redefines the original problem so as to be able to solve it adopting its own standardized and conservative schema. The corporation colonizes the problem. In the best of cases, the situation, as we are trying to demonstrate, can escape control due to the growing inadequacy of the self-feeding solutions that the corporations propose. Therefore, problems autonomously raised by the citizens also occur. Yet the perception of the problems, however spontaneous and "real," is never completely separated from the professional-technical-scientific discourse which, today, prevails over the ethical-political one. Problems are (im)posed by technicians instead of being set by the peoples' representatives. If this is the general trend, nonetheless, sometimes it may become possible to escape corporations' control, especially when their solutions become more and more inadequate.

After having been operating for some time, the organization no longer pursues its original goal. The new real goal becomes to keep the organization existing, working and possibly growing. In sociological terms, we can say that the function prevails over the goal. If the function has become the real goal, it means that the organization is pursuing something it has not been designed and established for. Because the organization mainly strives to outlive and grow, its goals and the means to pursue

[3]Italics in the original.

[4]I will describe how in practice this behavioral pattern develops in the following Chaps. 5–7.

them shift from being merely technical to being substantially political.[5] Needless to say that nobody in the organization would ever admit this.

Could Citizen Participation Be a Solution?

Some claim that deliberative democracy and direct citizens' participation may be the cure for this misleading situation. If "incompetent" citizens were left free to express their needs and preferences, they may favor environmental policies and social change. In fact, most of the time, citizens are part of organized groups of vested interests and, in other circumstances, the same citizens act as autonomous individuals. Citizens as members of vested interests will pursue their organization's goals, i.e. they would benefit from the functioning of the system, independently from the real goals. When the same citizens act and get together individually – without the mediation of the organization – they are keener to single out different goals that are likely more direct and authentic. This does not happen all the time and certainly not so automatically as it has been described for many reasons. To express diverse preferences, citizens (a) should be properly informed about non-conventional solutions to current problems; (b) should be able to organize politically and stand up for innovative solution implementation. But people tend to rely on the most powerful institutions that are likely to be the established ones. Thus, if vested interests and organizations have proved more effective than citizens' democratic institutions, citizens will prefer to give their support to their professional organizations rather than devote themselves to aggregate dispersed and diverse people around new problems that are not even yet well defined. Our knowledge of reality tends to be opportunistic[6] and so it happens when we decide to support one or another project or political strategy. The strategies that have been already applied and the old projects, although ineffective, have an advantage over the most innovative ones. This situation can be modified if we introduce an ideology, or a more comprehensive discourse.

Ideologies, according to Marx, are rigid schemes of thought. They reproduce the falsities that are necessary to allow the dominant classes to control the capitalist exploitation system. Therefore, ideologies are the opposite of critical thinking. However, ideologies are also defined, in non-Marxist terms, as a system of shared social beliefs. If we accept this second definition, ideologies can be useful in "creating time" for decision-making. In fact, they describe development in a series of predefined steps or phases. Even though the evolution described by ideologies may not be scientifically proved – as in the past somebody naively claimed[7] – the

[5]Large, and even mid-size, engineering firms typically invest a lot of time and money in lobbying.

[6]See Myrdal's introducing quotation of this chapter (1975). *A propos*, it is interesting to note how Paul Feyerabend reports that Thomas Kuhn, in writing his "*The Structure of Scientific Revolutions*", spent a lot of time interviewing scientists, which is certainly an uncommon research method in philosophy. The rationale in Kuhn's approach is clearly that even scientists choose their researches on the basis of considerations that often have little to do with a genuine lust for truth or knowledge.

[7]See, e.g., Walter Rostow, quoted in Chap. 2.

perspective of a predetermined development path may make people more willing to accept some inconveniences since these are perceived as merely temporary obstacles on the way to progress. For instance, if people would rely on the "ideology" of Rostow's development stages, they may be more disposed to accept some sacrifices during the "accumulation" phase – being convinced they are temporary – while still hoping for a future growth. A similar point may be made, from a political perspective, regarding the dictatorship of the proletariat – a period, considered necessary, during which personal and political rights may be restrained – that would likely open up to a fair communist society. In the case of a traffic revolution, if it is supported by a shared ideology, people may assume that by reducing traffic supply you would encourage the establishment of neighborhood shops, of a housing and job market that would favor a propinquity between work and residence, of pedestrian areas that would reduce dependency on driving, and the like. People would make their choices according to these expectations even if, in the short run, they may be penalized by lack of good services and still forced to commute. Moreover, if they expect a certain development, but the evolution proceeds too slowly, they would be encouraged to protest actively to hasten the evolution they have in mind. Even more practically, if I expect that public transportation will be improved, I'd be more willing to accept a temporary discomfort traveling on shanty trains instead of immediately buying a new car. If I assume that the most viable transportation policy must be based on public transit rather than on new throughways, I will not advocate for having new roads and will likely protest with the decision-makers for building new trains.

If a shared ideology is missing – i.e. a scheme that describes both the future desired situation and the subsequent stages to get from here to there – the appeal for a new self-organization and direct citizen participation diminishes considerably.

Environmentalist Research and Training: Old Wisdom, Current Insanity

In some cases, problems are put forward by citizens who are challenged by new occurrences, instead of being theoretically formulated in the restricted boundaries of the industrial-professional community. However, administrations are only capable of responding with "specialized offices" even to new situations. These specialized offices have corresponding "Departments" in the Universities where the personnel are trained. In this way, problems return to being filed away in departments overseen by chemists, biologists, physicists, natural scientists, etc. The process happens to be perverse especially regarding environmental issues which are by nature an all-inclusive subject. The customary approach to administration would even have its own merit, as long as it does not remain rigid and unable to adapt to change. It becomes ineffective when new problems arise that are not easily classified in the old specializations created to face problems of a completely different nature.

It should be the task of the university, with the advances of research and culture, to change creatively the way we identify and manage environmental problems.

In reality, this does not happen, because schools respond passively, with few exceptions, on the two fronts that concern them. The first front is that of research and creativity: safeguarding traditionally correct academic approaches prevails over the testing of new ones. The second front is academic training, which guarantees administrative and professional personnel a traditional education, without entering into a critical dialectic with practitioners. The vicious circle gains strength, and the corporate-technical-scientific powers-that-be take advantage of the situation.[8] The professional corporations give their members a feeling of security and identity. This structure has favored great progress in the past because of the standardization of knowledge, professional competence, and practicality. However, we cannot overlook the fact that, from approximately the eighteenth century until a few decades ago, scientists and scholars have been picking up problems never considered before and it was undoubtedly revolutionary.[9] Now this science and academic structure seems clearly obsolete, considering its clamorous and repeated failures both regarding urban organization and mainly environmental problems. Above all, the problem-solving structure is supported by people who lack the desire to produce revolutionary ideas that should characterize many intellectuals and progressive leaders.[10]

"*Offelè fa' el to mestè*" translated from the Lombard language means "pastry chef, do your own job". It is an old saying used to reproach those people who want to do more than what they are requested to do or are formally specialized in. Indeed, we often praise the old wisdom of the ones who utter: "If each one would do their own job properly, the world would be immediately better!"[11] It was probably true in

[8]The juxtaposition of the word "department" with the word "university" might appear almost ironic and an oxymoron. Defining the academy as a group of "Departments" is etymologically opposed to the "university of knowledge," which by its nature would be difficult to compartmentalize. We must take seriously the condition in which the role of the Departments and of the specialist sectors grew to the detriment of the Colleges and of the University in its entirety. I will return to this later when arguing about education of traffic planners.

[9]"Great authors, scientists and philosophers since the seventeenth century … saw things never seen before, thought thoughts never thought before …" (Arendt 1998, chapter 6).

[10]"*The academic world is in love with disciplines, departments, fields, programs, and schools. To some extent this is beneficial. Serious thought requires distinctions. Analytic methods rely on dividing and conquering the material under examination. Disciplines provide resources for these activities through their canons, methodologies, and vocabularies. They shape the questions that are asked, and help to define what counts as answers. Even when this love affair is not good it is often relatively harmless. Identifying with a discipline or department is like belonging to a guild or a union, or cheering for the home team. Some people like uniforms more than other people, but this is not a matter of great concern.*

However to some extent, this love of academic order is pernicious. It can be more like sophisticated (but still vicious) forms of tribalism than conformity to sensible epistemological canons. The love of institutional distinctions can distract or deflect us from the most urgent problems we face. The world's problems do not respect the order of battle imposed by university administrations and policed by college professors, graduate students, and professional organizations. They come to us in their own terms. If we want to make progress on at least some problems, we must follow where they lead, rather than attempting to impose on them our favored categories of thought." (Jamieson 2010)

[11]It goes without saying that this aphorism reports a very conservative and authoritarian bias. In fact, it does not presume any possible change, modification of social roles, nor class mobility.

modern era, but now this approach looks very outdated and is the real cause of several environmental problems. Actually, we have all learnt to do just our own job – which is good – but we have completely forgotten about what the others do and can do – and this is bad. We have taken for granted that society automatisms – e.g. the professional job market – would have balanced demand and supply also in the field of professional competence. Because we are poorly informed about what others are supposed to do, hence we do not actively look for somebody who can solve our problems, but we passively "buy" what specialized problems-solvers try to sell us. We comply with a supply-led market and find it hard to express our own demand. For a long time we have been neglecting the links among professions and skills, assuming that a successful outcome would have come from the sum of different parts. A new wisdom, suitable to the contemporary situation, would pick up an opposite view and acknowledge that we should all know at least something about what others do or can do, and relate to them.[12]

I mentioned that traffic plans are the hunting grounds of civil engineers. Waste management belongs to chemical engineers, even if they currently prefer to be labeled as "environmental engineers" or "environmental chemists." Urban planning was in the firm hands of the architects, but now they are in competition with urban planners, many of whom are still trained in architectural departments.[13] Although architects are trained to take a comprehensive view of problems, they still have a design approach, even when they deal with comprehensive social matters. They have a propensity to focus on the final product – the mythical "plan" – and overlook

[12]Fritjof Capra (2007) has recently proposed a biography of Leonardo da Vinci and an analysis of his unspoken scientific method. Capra claims that, if Leonardo's approach to scientific procedure had prevailed over Galileo's and Descartes' Method, we would have now a different science. It is interesting to note that the Latin word "*methodus*", of Greek origin, is composed by "meta" (beyond, above) and "odòs" (path) and thus it refers to something that connects all parts of knowledge. We need a method to unify disjointed notions. Paradoxically, Descartes' Method has not produced a coordinated knowledge, but rather its opposite. In fact, all pieces of knowledge became coherent thanks to the method which was able to unify them. The unity was only theoretical, though. As soon as the theory was questioned because it was becoming ineffective to solve new problems, the unity was necessarily transformed into an ideology and into a dominating scientific discourse. My philosopher friend Alessandro Tessari pointed out to me that, the word and the notion of "method" being philosophically very sophisticated, they were rarely used by the Romans (except occasionally by Vitruvius and by a fourth century A.D. poet, Claudianus, who referred to something very practical) who had no interest in – and somehow despised – theoretical thinking. The word and the concept of method regained the limelight with Descartes' "*Discours de la méthode*". Romans preferred to use the word "*ars*" akin to the idea of skills, craftsmanship. Nonetheless, Tessari claims that Descartes drew ideas and concepts from the Catalan thirteenth century philosopher Ramon Llull and by his followers, as it is proved in his private and public correspondence. Llull anticipated Descartes' attempt to provide a method for universal knowledge, except that his attempt was identified with Latin names such as "*ars*", "*clavis*" or "*mathesis universalis*" and so it generated some confusion. In fact, Llull studied Kabala and searched for the "*araba phoenix*", i.e. the cornerstone of all knowledge.

[13]I mainly refer to continental Europe and particularly to Italy and France. In the US there is a clearer separation between architects and planners, the latter being more open to social approaches than to physical planning.

the political and social process of the plan implementation. In architecture schools – now more than ever – students are taught the solutions to problems, rather than the framing of solutions or the detection of new problems.[14] A conservative attitude dominates and the radical thought which forms the base for innovative research and for the creation of new ideas, is marginalized, if not truly disdained. Architects and planners are urged to present immediate outcomes and do not "waste" time in speaking and arguing. This kind of training is designed to be specifically practical, but it proves unsuccessful. In practice, however, when technicians come across a known problem, the immediate desire to apply the familiar solution, without any further reflection, is the result of their instruction and attitude. This attitude, which is not efficient at all especially in current times, is even more likely to be adopted because it is simply the easiest.

"A Patch Is Worse Than a Scratch": Two Reactions to Contrast Current Ineffectiveness

With the growing complexity of contemporary society, this attitude leads the communities of technicians, and thus their corporations, to react in two ways: (a) to subdivide internally; (b) to turn to interdisciplinary collaborations.

First reaction: (a) internal discipline subdivisions. Some readers might be familiar with the "polluter pays" principle. According to this principle, a price is established by which a polluter can pay in order to have the right to pollute. At a conference, one lawyer lecturer, specialized in tort law, not very well-read on this specific environmental policy subject, made a well-documented and elaborate presentation. Unfortunately, he described all of the possible ways of prosecuting polluters, in order to "make them pay!", which in his perceptions meant "to punish them" instead of forcing them "to refund" for the damage produced as the "polluter pays" principle actually imposes. After the conference we tried to explain to him where he went wrong, but no one was successful. Consequently, in order to avoid further embarrassment to the organizers of the event and the presenter himself, we decided to compliment him for his contribution, which was, in fact, quite well done, although absolutely irrelevant and out of place.

The guild's first reaction to the loss of ability to problem-solve is internal specialization. Internal specialization means that professionals with a technical (engineering) education take on the role of sociologists, jurists, scholars in communication and public opinion, economists, or moral philosophers. The result is that the core of the project stays anchored in the main disciplinary background

[14] I owe this idea to the architect Paolo Soleri with whom I recently had a conversation in Arcosanti, Arizona. Arcosanti is a utopian village/community on the road from Phoenix to Flagstaff. The experiment is very interesting and successful. However, the participants do not consider any possible political promotion of it.

and *forma mentis* of the technicians. Other analyses – which meanwhile have become necessary or are requested by citizens – are somehow carried on by the technicians who are part of the guild, but, unavoidably, in an amateur style. Architects and urban planners generally call for a comprehensive approach to urban and environmental problems, and they are particularly keen on this approach because of the epistemic basis of their discipline, which implies a general view of the situation and the coordination of actions. Nevertheless, the point of departure remains technical and physical. Architects are educated as project designers, but planning is a category of social change and would be better treated by sociologists, political scientists and humanists. The historic inheritance and the scarce dynamism of some academic institutions prevent new ways of thinking. Obviously exceptions exist, but we have not yet seen a cultural revolution that has allowed humanists to take over the governance of the territory and lead technicians. Actually, it is not that the one is better than another. It is just a matter of efficient organization of competence. Too rapid a change could cause disastrous consequences. We cannot get rid of engineers and technicians immediately, no matter how inefficient they are. Nonetheless, we need to begin thinking that some problems are not merely technical ones and that new skills might be necessary to tackle them. These new skills should be created in social science, policy studies and humanities rather than in technical and science schools. As Feyerabend explained, if the Church had immediately permitted Galileo to put his own convictions into circulation, while they were still confused and unproven, the whole intellectual and power structure existing at the time would have collapsed. This did occur eventually, but in a gradual manner which took centuries and a few execrable bonfires for which only recently the Church apologized.

A reaction to the complexity of problems, based on the internal divisions of the guild's competence, does not permit a rapid progression towards a radical solution. Above all, it does not change the identification and framing of new problems, and it preserves the disciplinary structure. Competence in "other" subjects rather than the core corporation's business, is rarely as sound and systematic. Moreover, in day-to-day operations this competence cannot even be excessively publicized. In fact, in the crucial phase of applying for an assignment, a company proves reliable if it is competent in the guild's core subjects rather than in side competences. Thus, although professionals may consider these side competences as priorities, it is not prudent to emphasize them if they want to get the job. The non-specialization outside the traditional competence of the professional guild is not a completely negative feature. Practitioners from different disciplines who make incursions, even as amateurs, into the territory of others can bring an unexpected creativity. For example, the architects and urban planners who organize civic participation may introduce elements of illogic and confusion, but will simultaneously break the rigid corporatist schema structured by sociologists and academic political scientists. In addition, the lack of disciplinary rigor also introduces a sense of confusion, a feeling that "everything is possible" into the rigid corporatist discipline. Eventually, in order to win an audience, you must be part of the guild. But being part of it weakens the revolutionary potential. So we are trapped. This confusion would be creative and helpful if, and only if, it could break the monopoly of the traditional methods of the guild. Instead, being "amateur" the architects and the engineers recycled as sociologists or communicators are doomed to play a marginal role. Moreover, on the basis of my decades long experience, I realized that the ones who engineering companies put in charge of the so-called minor issues – like citizen participation, communication, environmental education and other "trash" like these – are the people who had proved least successful as technicians.

Second reaction: (b) multidisciplinary collaboration. The case of multidisciplinary collaboration is similar. The leadership won by a specific corporation – such as the civil engineers in traffic plans – implies that the very identification of the problems and their framing are so rigid that the multidisciplinary approach cannot substantially change the way they are treated. The head of the team formulates questions that others on the team are only supposed to answer, rather than to reformulate. And we know that questions contain their answers.

In both cases we are looking at a situation that cannot enact any real improvement because it is compromised from the logical-operative point of view. The professional guild cannot accept solutions which lead either towards the lessening, or the loss, of their own role. Corporations and technicians have a vested interest in resolving a problem temporarily, even when it would be possible to eliminate it completely, at least in its current manifestation. Is it likely that a university researcher specializing in structural engineering could declare that bridges, underpasses, and viaducts are not necessary anymore? Or that they are needed in lesser quantities? It is improbable, unless we meet with people who want to revolutionize both their own existence, and the discipline they studied so hard. This cultural revolution would have a chance if it were founded on the true intellectual desire of a few scholars to explore new possibilities. But if the same university researchers are also professionals aiming at being hired by some local or central government to design a plan, then the probabilities of revolution become minimal. The possibility of changing the approach approximates to zero when politics does not express any autonomous direction from the power of the corporations, and rather is a slave to them.

We idly accept the premise that Universities are to prepare students to compete in the job market. But an opposite point of view is also possible, namely that Universities should be autonomous from the business community in order to be able to propose innovative ideas freely and to provide the skills to apply them in everyday life. It would be the role of entrepreneurs to make some of those skills profitable in the market. Universities should graduate students who can be resources to the administrators and entrepreneurs who will employ them. The new competencies which would be created by a liberal culture that exercises free thought will forge new ways of confronting the problems of society. Only with this approach will Universities be able to take a cultural lead in the contemporary world. Autonomy from politics of science, education and research has always been obtained thanks to a political act of the authorities subsequent to a rebellion of scholars who wanted to affirm their ideas. As an example, let me quote once more Arendt reporting about the Royal Society. When the Royal Society was founded, "members had to agree to take no part in matters outside the terms of reference given by the King, especially to take no part in political or religious strife. One is tempted to conclude that the modern scientific ideal of "objectivity" was born here, which would suggest that its origin is political and not scientific. Furthermore, it seems noteworthy that the scientists found it necessary from the beginning to organize themselves into a society (a guild, to use the terms of this essay) …". "But an organization … is always a political institution; where men organize to act and acquire power" (1998: 271n). Arendt also quotes Whitehead remarking that it is "no accident that an age of science has developed into an age of organization. Organized thought is the basis of organized action". The problem with organized thought and with organizations in general is that they are not likely to reform themselves on their own impulse unless they are vigorously challenged from outside.

Morality and Traffic Professions

The Problem of Technical-Scientific Environmentalism

Another problem deriving from intellectual disorder and professional corporatization relates to certain dogmas passively adopted by the majority of militant environmentalists. Also in this case, we notice the dramatic result of a syndrome that emphasizes the passive and trusting application of a solution around which the problem is formulated. The process should evidently be the other way around: first define the problem, and then study the solution. Surprisingly, this sequence is not so obvious. The less you are biased by referring to previous solutions, the more open minded and innovative you have a chance to be.

Specifically, the illogical approach of letting the solution precede the formulation of the problem depends on corporations' and lobbies' power. In general, it is a consequence of contemporary science and technology politics. The sustainable development-related policies have reinforced this illogical approach because governments prefer funding projects that are immediately implemented rather than financing alternative research and the elaboration of long term plans which should be the real role of public and research institutions. By acting so, research and public institutions lose in autonomy and political/intellectual power. A good example of this misleading approach is the (so-called) research, strongly encouraged by EU, on the collection of "best practices". When they need to cope with a new situation or problem, European local administrations are addressed to catalogues of "best practices", namely to projects that have been already successfully applied by other administrations. If they choose – or refer to in the application for funds – one of those previously adopted practices/solutions, they are more likely to obtain funds. There is nothing wrong with encouraging administrations to communicate with others and with seeing and re-elaborating what has already been done. However, another policy is possible, specifically to fund new plans and new improved visions. Unfortunately, this is not popular nowadays and there are no funds available. The unquestioned "best practices" method is just another aspect of the conservative and lobby-dominated approach to traffic policy.[15]

Environmentalists are also responsible for not questioning the corporatist approach with enough conviction. Unfortunately, the corporatization of society affects also the environmentalist organizations.[16] Political discourse remains anchored in a ground dominated by those who created the environmental problems. The jargon shared by the majority of the population is the language created by technicians and scientists. Consequently, political activists find it difficult to move the battle from the usual field of technical operations to an alternative ethical and epistemological approach. Only the latter would allow serious progress towards environmental education and the resolution of contemporary city problems. Many environmentalists channel their energy into trying to convince industrialists that with new, low impact, low-pollution technologies it is still possible to make good profits! Environmentalists look impatient to see some immediate result of their action. This attitude proves that they are insecure: dramatizing every situation prevents them from elaborating long term vision and policies.[17] Once again it is worth comparing the environmental language with that of the worker's movements. Labor leaders normally use rhetoric and abstract concepts that nonetheless the audience easily understand and link to the current political discourse. They can safely call for a generic theoretical "social justice" or no better specified "workers' freedom and dignity" without the risk of being ridiculed as wordy and unpractical radical-chic dreamers. On the contrary this would certainly happen if some environmentalist emphatically spoke about, say, "human right to health" or the "right of Nature to

[15] A book is like a key. Keys can open doors in as much as books should open minds. However, keys are used also to close doors in as much as books may imprison minds. We are inclined to think that an open mind is good and to imprison the mind in fixed schemes and biases is something to be avoided. It is not so all the times. Most intellectuals cannot resist the appeal to know, to reinvent, to elaborate new ideas. Pioneers of innovative thinking are generally more loved and admired – although usually less powerful – than the gray catalogers focused on details and rigor. But new ideas and alternative interpretations, no matter how fascinating, are often vague and flawed. Nonetheless, gray catalogers – the ones who codify, make clear and procrastinate the use of shared concepts, in which they lock the lazier thinkers and who are drawn to technique rather than to philosophy – operate in the spirit of the ethics of conservation and short term efficiency. Thus, books, the keys of knowledge, are necessary either to open the minds or to lock them by binding students and technicians to a few certainties that provide psychological self-confidence. Assumed that books work both to open and to lock, I dare to claim that for some decades we have given too much importance and prestige to catalogers than to innovators. Maybe, it has always been like this: in order to assert new ideas – if they are really new and keen to change the world – it is not enough to compare them with the old ones: one ought to fight in order to be listened to and to win the right to promote them. I am not claiming that all new ideas are good per se and new thinking is always pure. I just maintain that it is necessary nowadays as time has come to unhinge old paradigms.

[16] According to the classification I proposed in chapter two, I should better refer to organizations dealing with environmental issues, having defined environmentalism as something more specific.

[17] See again chapter two about the "sustainable development" fraud and the problem of the missing environmentalist ideology.

integrity", instead of bringing some questionable objective data on pollution-related diseases or fake (but quantitative) information on exposure to some cryptic microwaves gauge.

The Need for Radical Thinking

If we do not esteem radical thinking even in the knowledge citadels, change becomes improbable and we are doomed to stay imprisoned in the crisis of contemporary conformist environmentalism. George Bernard Shaw argued that the wise man is the one who gets adapted to the world, while the madman will try to shape the world according to his desires. As a consequence, progress cannot be brought but by madmen! We also like to quote Benedetto Croce who wrote that "... people as well as individuals, if they do not progress, or strive to go further, they go backward; they cannot stay still and calm without getting corrupted". This is true also for environmentalism ... and traffic policies.

Most people have a distorted perception of the most dramatic contemporary urban problems since they apply old definitions that thwart decision-makers from dealing properly with them. E.g., almost all the surveys conducted in affluent European cities prove that both people and decision-makers list urban micro-criminality and traffic at the first positions among the most serious city problems. These problems are perceived as the cause of most of the other problems. In reality, things are not completely like this or, at least, this misleading perception is one of the major impediments to solving them. The situation is complicated by the fact that citizens, authorities and scholars – who, by the way, bear the major responsibility for continuing these biases as they should be the ultimate prejudice hunters – feel hopeless and react by showing an arrogance which is both out of place and a proof of a lack of confidence. Everyone assumes that the solution is well known and they trust the specialists to define the details. Nonetheless, nobody really inquires in depth what the problem is, since they dare not invest a minute in critically examining it or asking whether there are ideologies which are suited to locate it in a broader hermeneutic frame. Paradoxically, decision-makers, scholars and citizens are absolutely right believing that they know the only solution available. But this is the real problem, i.e. the lack of any intellectual approach to traffic problems: people refuse to think that "there is nothing more practical than a good theory" in order to deal realistically with complex problems. Presuming to know the "solution" prevents people from being interested in studying the problem in depth. In this way we spell out the paradox: we apply well-known and tested – sometimes cleverly refined – solutions to problems we inaccurately know or to issues of which we have an outdated, misleading, wrong, or superficial knowledge.

What I have called "solutions", in a philosophically correct language, should be defined as "techniques". They can be either simple century-old tools such as road construction, urban rails for mass transit, or cameras and police services to fight against criminality. Technical innovation develops inside the constrained boundaries of this old technology. Because easy-to-understand technical innovation still attracts the childish attention of many, most people become focus on this nicely packaged solutions and are distracted from the real problem. Thus we have confused science with technique also in dealing with urban problems. If we want to progress we need to change our way of thinking. Unfortunately,[18] a cultural revolution can hardly be born in the mind of a single enlightened scholar, especially if we are speaking of social innovation. Technological and philosophical revolutions require hard fighting against powerful academies and corporations that do not have any intention of easily relinquishing their power.

To fight against the power of the conservative guilds and academies, we probably should substitute "competent professionals" with "professional incompetents" – following a provocative suggestion that Feyerabend proposed 30 years ago. The latter could be able to see new problems (or the same problems from a different perspective) although they do not yet know the solution. The outcome would mean more creativity both in the framing of the problem and *then* coming across the solution.

As a matter of fact, to be able to resolve urban traffic problems, we need to change the problems! ... At least to reformulate them dramatically. This approach – which should be fascinating for a scholar – is not a viable one at present. The past and its dominating structures are still sturdy and it is not likely to forecast an immediate cultural and technological revolution. However, there is nothing else we can do, which is why we need to make it clear that the building of traffic infrastructure is principally an ethical problem related to both environmental ethics and to the responsibility of professionals.[19]

Panic and Compliance

In traffic related issues, politicians and practitioners have often failed to demonstrate foresight. Most of the time they behave like the man, whose house being on fire, casts himself directly into the blaze because he is afraid he will not survive. By adopting this mindless strategy, we have destroyed much of the existing preserved land in order to enlarge roads; we have cut trees and asphalted parks to build parking lots and garages; and we have ruined beautiful views to build useless infrastructures

[18] Or maybe fortunately!

[19] Besides the basic readings of Kuhn and Feyerabend that have inspired the general approach of my essay, I drew more than an inspiration also from Bruno Latour's more recent essays (2004, 2005) and in his classic *Science in Action* (1987). Precisely, when I propose how to plan urban traffic, the reader will find some echo of Latour's "practical metaphysics" and of his claim for the necessity to consider the plurality of worlds. In the introduction to my essay on "Città flessibili" (Flexible Cities) (2009) I extensively argue on Utopias and claim that a contemporary utopia should be the transformation of Utopia from an island into a continent so that a major issue becomes the tolerance and the connections among different utopias. The essay will soon be available in English. Hopefully.

which have created more and more traffic. People who behave in this way call themselves "practical". The apparent situation of traffic all over the world provides sufficient evidence of how unpractical the actions of self-appointed practical persons really are. The only gray matter they are able to employ is cement.

As a matter of fact, we cannot exaggerate in the opposite direction, as if, for example, when the house is on fire, we do not stop painting the walls, and continue to worry whether beige would have been better than a light blue. This attitude is as stupid as the previous one. Both mind-sets are inadequate when it comes to managing traffic. A correct strategy should keep management rigorously separated from planning. Only in this way is it possible to conduct research and to find alternative approaches – without renouncing the desire to cope with everyday problems.

This approach would also guarantee a necessary gradualism for change that would include some concern for the psychological and economic protection of professionals and businessmen trained in the traditional approach. It is unacceptable that we cannot stop for a moment to reconsider urban problems deeply, and only afterwards think of appropriate solutions. To be rational means – and has always meant – the construction of new realities. We could also introduce a rational discourse into the study of perceptions and behaviors. A free and creative policy requires some intellectual radicalism. However, it is not necessary to apply all the radical ideas simultaneously, but some of them could be introduced now and then to create some special situation. Or, we can design test areas in which all of the suggestions can be experimented with at the same time.

If we want to effectively approach mobility uneasiness, we need to consider that transportation, mobility, and traffic are complex problems – the characteristics of which are interconnected with many other social and technical issues. As Luhmann alleges with some mockery, "a single problem is not problematic at all". The practical solution seems to tackle each problem at a time. The outcome of this solution is the current disaster. We wonder how whoever is in charge of the mobility problem can define themselves as practical. As a matter of fact, these self-appointed practical people refuse to see reality or to prove anything else but their impudence. These so-called practical men have not solved a single urban problem in the last 50 years!

The "Banality of Evil", Ethics and Professions

The building of traffic infrastructure is also an ethical problem related to the responsibility of the professionals involved. The colonization of the specific problems by single professional guilds encourages professionals to focus on solutions rather than analyses. For instance, it has been a quarter of a century since personal computers and the Internet have become part of daily life, but they have not yet been located at the core of the analyses and solutions of traffic problems. When an expert on traffic is needed, a civil engineer is "automatically" hired. Very often this engineer is somehow connected with constructors and with companies interested in selling public transportation means. But managements of large transportation infrastructure companies operate mostly in the financial market and as political lobbies. The physical construction is delegated to other smaller companies controlled by holdings or somehow involved in political lobbying for getting contracts. Therefore, the decision-making process is partitioned in a number of independently acting bodies and there is no apparent responsible for the whole process.[20] Speaking confidentially, it is not difficult to persuade decision-makers of the necessity to tackle traffic in a completely different way. But eventually, decision-makers, professionals and citizens are overwhelmed by the famed Hamlet doubt: "… Who would fardels bear/ to grunt and sweat under a weary life/but that the dread of something after death,/the undiscover'd country, from whose bourn/no traveler returns, puzzles the will,/and makes us rather bear those ills we have/than fly to others that we know not of?[21]" To put it simply, we are afraid of new disasters, so we prefer the old ones to which we have grown accustomed. In the specific, Hamlet's doubt is reinforced by an unequivocal personal and short term convenience to act in a conservative way.

Professionals' behavior in traffic and environmental policies often prove what Hannah Arendt defined as "banality of evil". In a famous and controversial report, Arendt described Albert Eichmann's trial in an Israeli court for crimes against humanity, namely for having organized the genocide of an astonishing number of Jews during World War Two. In dealing with the Holocaust, the author went beyond the extermination camps and argued that the personal Nazi criminals' guilt was indeed a little thing in relation to the immensity of the crime. To defend themselves, most Nazi criminals claimed that they "just obeyed the orders". Arendt considered these people as nothing more than mediocre employees interested only in their career, despite their need to perpetrate diabolical crimes in order to claw their way up the career ladder. The real guilt of the organizers of the genocide was not the enormity of the crime, but it was their intellectual and ethical mediocrity. In a broader philosophical approach, Arendt argues that, because modern society operates by segmenting knowledge and tasks, producing an absolute evil becomes easy and banal. Even a collective crime, as enormous as the Holocaust, can be produced by ordinary people whose responsibility is, in their eyes, indeed fairly small. Most of the Nazi criminals, and specifically Eichmann, were not demons – as they are often described – but just ordinary people. Between the Holocaust and the

[20]To explain how people can avoid responsibility let me report the famous story by Oxford bioethicist Jonathan Glover: "Suppose a village that contains 100 unharmed tribesmen eating their lunch. 100 hungry armed bandits descend on the village and each bandit at gun-point takes one tribesman's lunch and eats it. The bandits then go off, each one having done a discriminable amount of harm to a single tribesman. Next week, the bandits are tempted to do the same thing again, but they are troubled by new-found doubts about the morality of such a raid. Their doubts are put to rest by one of their number … They then raid the village, tie up the tribesmen, and look at their lunches. As expected, each bowl of food contains 100 baked beans … Instead of each bandit eating a single plateful as last week, each takes one bean from each plate. They leave after eating all the beans, pleased to have done no harm, as each has done no more than a sub-threshold harm to each person." (Quoted in Shrader-Frechette 1991: 70).

[21]W. Shakespeare, *Hamlet*, III, 76–82.

environmental disaster there are clearly many differences. The nature of evil is different, but the characteristic banality of modernity works in a similar way. No modern individual feels guilty enough about their minor daily decisions to worry about a severe punishment. Nonetheless, moral and intellectual mediocrity of each of us may produce a condition of incessant and untamable degeneration. Arendt argues that we should put ethical political action at the center of the human condition instead of surrendering to the pessimistic "sociological terror" as described by Niklas Luhmann, who describes an invincible System which utterly dominates humanity.[22]

In the last decades, arguing about professional ethics has become more and more frequent especially in relation to environmental concern and bioethics. In fact, the invasive impact of new technologies and the non-return effects of their application on nature implies a careful use of them. At the same time, scientists have transformed from people of culture into highly trained technicians often uninterested in a comprehensive philosophical understanding and in the long term consequences of what they are doing. Bioethics has not only become a prominent branch of philosophy, but its principles are commonly applied by ad hoc established Ethical Committees. Ethical Committees regularly assess, value and approve many medical doctors' decisions and pharmaceutical companies' products. Certainly, the intrusion – or the return – of ethicists into the scientific community has not been welcome by the scientists and has generated an ongoing conflict. The way that Ethical Committees are established and how they deliberate is vigorously questioned, but I want to highlight how we do not have the same sensitivity for nature and for technological risks that we show for human and personal rights. Hence, it seems bizarre to speak about professional ethics of engineers who "only" want to build a bridge or a highway compared to ethical concerns related to stem cells or bio-engineering. Nonetheless, the environmental ethics and the decision-making problems are much closer than they might appear at first glance. The disproportionate availability of energy and the consequent possibility to construct virtually everything everywhere regardless of any limit in modifying the natural environment and balance, and without caring about the all-encompassing social effects of the constructions, should imply a professional ethics arguing about the engineers' behavior.

---

[22] I have been studying Luhmann's sociology intensively during my post-doc years. My interest in his writings was not, as it happens most of the times, that I liked his depressing ideas. On the contrary, I was interested in finding a way to escape from the trap Luhmann built describing the "system dictatorship". In this respect Luhmann acted as Dewey's good pedagogue who entraps his students so that, if they want to escape, they must learn. Arendt, Myrdal and Crozier readings helped me to escape Luhmann's gloomy trap of pessimism. At this point of my essay, I probably need to justify my recurrent quotation of Arendt. It is a consequence of some classes I have been giving since 2004 at the University of Bergamo (Italy) on An environmentalist reading of Hannah Arendt's *Human Condition*. I find this reading useful since I can put together issues such as epistemology, an analysis of contemporary society and end up with a call for action. Arendt's *Human Condition* has seldom been studied and quoted by scholars in environmental studies, but I believe that we can extract interesting clues which would help in understanding the contemporary environmental crisis from a political and scientific point of view.

Conclusion

This departure from the subject of mobility is driving me on another route, which is definitely more important, noble and comprehensive, but that I do not dare to pursue as it would take me too far from my more modest goal to deal with traffic problems. Coming back to physical mobility and traffic problem, we need to admit that:

- the idea that traffic is a social phenomenon has been neglected
- traffic depends on the location and supply of services and, thus, it is an urban geographical problem
- if we really want to study solutions – instead of adopting the ones we already know – we need to broaden our analyses in diverse fields.

The environmental crisis affecting cities is exacerbated by out of control traffic which depends on an excessive mobility. It should be obvious – but unfortunately it is not – that we are dealing with an ideological and political problem. To better the situation it is necessary to change our customary way of thinking radically. We should develop a new approach in order to change the future perspectives. Because the future has not yet happened, we can still change it. This is what previsions are for. Trend is not destiny. If I expect that they will build new roads and the roads will give me a more rapid access to services in the future, I will consequently make choices that will hasten decision-makers to implement these plans as soon as possible. If, on the contrary, the perspective is that no more roads are to be built, I will look for different solutions that can range from the most conventional, such as rails and public transportation, to the most innovative, such as a dramatic reduction of the physical mobility, with the development of telecommunication and a decentralization of services.

Chapter 5
Traffic Planning Critique

People who wax enthusiastic over comprehensive planning always start from the implicit assumption of faultless planning. Never do they suspect that comprehensive planning may equal to comprehensive blundering

Gaetano Salvemini (1939).[1]

Abstract Today, politicians and planners distrust long term visions and rational planning. Instead, they rely on incremental approaches and piecemeal solutions. Politics is both what is currently missing and what we really need. In general, one of the goals of modern society has been to get rid of politics and substitute it with automatic self-regulation devices. Urban Mobility Plans should be designed in a completely different way from the current approaches. The difference would be in the planning process and who is to be in charge. As a consequence, choices of plan content could be quite different. Planning is a category of social change. A well-functioning political system should be able to make new opportunities available and open to more possible choices. Scholars and technicians must play a role in elaborating innovative solutions. However, because the commissioners of mobility plans are reluctant to use competent professionals from fields outside traditional engineering, there is little opportunity for economists, sociologists, communicators, traffic psychologists, and so on, to develop new skills in the field. One of the major flaws in urban traffic planning is how communicators are employed. In the planning process, we are no longer asking the following essential questions: (a) why do we need to move? (b) does an increase in mobility imply a certain progress? (c) what are the

[1]Wittingly, Salvemini goes on saying: "*Such optimism depends, perhaps, on the fact that they are sure that planning will never be planned against their own wishes, and, of course, that their own wishes are flawless. They forget that the Almighty Himself had to acknowledge, one fine day, that he had blundered in his planned creation: he repented having made man and corrected that blunder by the Flood*".

C. Poli, *Mobility and Environment: Humanists versus Engineers in Urban Policy and Professional Education*, DOI 10.1007/978-94-007-1220-1_5,
© Springer Science+Business Media B.V. 2011

interpretations of the context in which we operate? If we give answers to these questions, we could envisage more effective mobility policies, at the same time realizing that consensus is part of the effectiveness. The most practical part of the traffic policy is to understand why people need mobility, not how much they now travel, but these questions are not given priority.

Keywords Planning theory • Politics • Politicians • Planners • Technicians • Professional skills • Social change • Communication • Data processing

The Missing Politics

Among planning scholars it is common to quote an anecdote reported by David Lilienthal (1944: chapter 18). Lilienthal was the first director of the Tennessee Valley Authority, the American agency established as part of Franklin D. Roosevelt's New Deal to challenge the poverty of some southern regions by using planning methods. The story tells of how a foreigner asked a local man "how he could get to Jonesville; after long thought and unsuccessful attempts to explain the several turns that must be made, he said," …: "My friend, I tell you; if I were you, I wouldn't start from here". The anecdote intends to describe an attitude very common among traffic professionals, public administrators and managers. They can easily define their goal and even design an optimal future situation, but they forsake all that lies between the final design and the means to achieve it. They do not consider the implementation process important since it belongs to the domain of politics and thus lacks technical and scientific dignity. This belief is sustained by a conviction which is deeply rooted in the more or less conscious way of thinking of contemporary scientists and technicians: if the project is good and rational, it will bear an intrinsic capability of self-realization and self-legitimization as if it were a divine revelation. An unconscious idealism makes these people think that, if the project is good, it cannot help but be implemented and, in some cases, enforced. At the antipode of this belief there are the bureaucrats who do not care at all about the project and its goals because they are enchained by the formalism of their own procedures.

To understand this attitude, we need to place it in the history of the last 50 years, or better in the intellectual trends and cycles.[2] In the sixties and seventies – and, for the unavoidable inertia that characterizes intellectual change, even in part of the

[2]Albert Hirschman (1982) proposed an interesting social scientist and economist perspective of how people's behavior periodically shifts from political involvement to a more private attitude. Although, in principle, I am quite skeptical about the possibility to excerpt realistic trends in human behavior, in this case Hirschman's model helps. True that the model is hardly predictive and it explains *a posteriori* what has happened in the last 50 years, as most economic and social science models do. Nonetheless, it may give a clue of what it is going to happen in the future. Because Hirschman's predictive trend matches what I hope will happen in the coming years, I take it and strive to make it really happen, thus it is more likely to happen.

eighties – the idea that it was possible to apply new theories and ideologies to rationally and politically plan urban areas had spread in many intellectual and political circles. Courses of planning proliferated in higher education and hundreds of books and papers were published. People expected radical changes and were confident that progress would have fixed the several pitfalls provoked by the recent urban growth and by the impetuous radical change from a rural society into a completely industrialized one. Certainly, this catharsis did not occur, thus planning theory and practice fell into disgrace since people expected too much from it and, as an obvious consequence, were disappointed. The reaction was to distrust any long term vision and rational planning – not to speak of political involvement and ideological objectives – and instead rely on incremental approaches and piecemeal solutions.

As a consequence, politics is both what is currently missing and what we really need. Regrettably, in present days, politics is discredited. The fact that politics is disgraced is both intrinsic in the structure of modern society and is a specific contemporary occurrence. In general, one of the goals of modern society has been to get rid of politics and substitute it with automatic self-regulation devices, i.e. the supremacy of the system over the actors.[3] Obviously, this goal has never been fully achieved. Politics has always played a crucial role even in those societies that seemed the most corresponding to the ideal type of modernity. Therefore, at the same time, some advocates of politics have always been the ones who called for political action in government and in schools. Nowadays, the discredit of politics is probably at its highest. This is honestly quite strange, because it is exactly what we need to countervail the worsening of environmental and traffic problems. Politics is necessary for prompting change. In the previous chapter, I tried to explain why this apparently illogical attitude prevails. When people are insecure and panicked, they are not open to innovation and stick with the old beliefs which in this case were the ones representative of the *passé* modern society. The modern project intended to get rid of politics as it considered it a hindrance on the way to civilization. Insecurity and fear of change are the reasons why the current environmental discourse is still based on technicalities rather than on politics and ethics. People assume that a technical discourse and a presumably objective mechanism are both more neutral and safe than any political competition which is considered hopelessly unfair. The specific occurrence is sustainable development (fraudulent) ideology: its success has been possible thanks to the discredit of politics and the dominance of a technical environmental discourse.

David Lilienthal gives us another interesting clue. In the1930s many professionals and scholars wanted to visit the Tennessee Valley Authority offices to understand the secrets of its success. Eventually, after having listened to several explanations about the working methods, they asked Lilienthal to see the "Plan".

[3]Literature is abundant on this topic and I have reported some of it in the second chapter, mainly referring to Giddens (1991) and Baumann (1993). However, I owe most of these ideas to comparative readings of Crozier, Luhmann and (partly) Habermas about the competition between the system and the actor.

The director replied that there was no document called The Plan. The real Plan was in the method they were using. The process was more important than the material design of the forecast situation – which was indeed the result of an efficient and creative decision-making process, based on clear values and goals.

The Urban Mobility Plan (UMP) and the Mobility Planning Actors[4]

These considerations, with the ones described in the previous chapter, should justify why I claim that an UMP be designed in a completely different way from the current approaches. The difference, in substance, concerns the planning process and the people in charge of the UMP and only as a consequence does it regard the possible content of the plan. Thus, the content and the individual projects must be intended as the consequence/result of the planning process.

During a presentation of my planning procedure, as soon as I uttered the words reported in the previous sentence, I was verbally assaulted by an old and well known engineer who has designed dozens of traffic plans, and who was employed by a big company. First year students in planning are immediately taught the difference between plan, planning and projects. The engineer responsible for dozens of inefficient plans was clearly confusing the plan with *his* project. He was also afraid that, by acknowledging the difference, he would have jeopardized his lucrative activity and his power. When a professional is working for a company that already has a

[4]I report on what currently happens in most local and central governments. This essay's main goal is to explain why the exceptions are so infrequent. This is the result of two factors: (a) the structure of the decision-making system and the organization of lobbies and guilds is firmly established; (b) innovative politicians and scholars have lost hope in being able to change their approach. Planning scholars have abundantly elaborated on planning theory especially in the years between 1960 and 1980. Thus, in the following paragraphs, I am not adding anything that has not yet already been said about planning theory. Rather I mean to report about my long experience as a consultant and a member of local and central governments. I am confident in saying that theoretical elaboration did not affect practice in as much as a scholar might have expected. It is likely that theories were not good enough to be applied. The reaction has been that scholars have refused to elaborate new more suitable theories and have given up in their efforts to change. The same story applies to comprehensive planning and to any planning in general. Scholars and practitioners have lost hope and have accepted the situation as immutable. The good news is that already 30 years have passed since this intellectual depression began. It is good news because it is scientifically proven and tested that every 30 years the intellectual mood changes and we can expect an inversion of the cycle which will make theoretical elaboration and long term planning popular again. Joking aside, I present a general scheme of how traffic infrastructures and plans are realized. I have analyzed if in different countries and situations the decision-making process and the content of the decisions could have been substantially different. The conclusion has been that this structure is so entrenched in technology, economics, and politics that there are virtually no real exceptions. This does not mean that we should accept traffic congestion in our cities nor that we should give up questioning the concentration of power in constructors' hands.

solution to traffic problems in its tool-box, there will be no plan. The solution will coincide with the goal, the project with the plan.

The mobility planner should be a person who can think as a political and social scientist, who knows about the communication and decision-making processes. Because planning is part of the social change process, the planners should consider the objective situation in which they are going to operate. Emotions, the hidden goals of the diverse actors, people's perceptions and so on, are all part of the objective situation and should be treated as such with the specific competence required.

In planning, depending on the method or the technique adopted, we can either establish goals *ex ante*, or identify them during the planning process itself. Currently, instead, all too often the definition of the general goals is kept implicit in the assignment given to the planners by the elected representatives. Thus, the representatives bypass the planner's analysis of the citizens' preferences, which are presumed or identified by the techno-political mandate provided to the planner. Often politicians possess an instinctual skill in understanding what the majority of the people want, and in this respect, they might be able to manage most of the situations that the planning process requires in both the choice of general goals, and during the subsequent negotiations. More rarely have they undergone systematic training in decision-making processes analysis and usually they do not know what other sources of knowledge they might use to produce a better plan. By a better plan, we mean one that can be implemented more easily by first seeking citizens' consensus on its content, and then evolving to become the output of an already accepted process. The lack of a systematic approach in the planning and decision-making process inhibits change both from the perspective of the specific choices and of the procedures adopted. Consequently, the lack of an approach based on well-structured thinking reinforces a conservative attitude, which results in a trivial outcome. An independent, rational theory would more likely produce the innovations required. Sometimes change is required to fulfill needs that are not yet precisely expressed by groups of people. A well-functioning political system would be able to delineate and make available new opportunities and more possible choices. Scholars and technicians should play a role in elaborating new ideas, innovative solutions, and eventually should be encouraged to formulate plans through institutional organizations.

Decision-makers should hire specialists in transportation planning (or in general planning) with training more appropriate to the contemporary social and political situation. Politicians are accustomed to hiring highly qualified communicators for their election campaigns. In the last few decades they have realized that their belief in their own instinctive sensitivity in understanding people's preferences is not enough in a society in which communication is the crucial issue, and personal contact has been substituted by a number of new and old media. The decision-makers' lack of consciousness about the possibilities of a more communication-oriented mobility plan makes it easy for the engineering community to allow these decision-makers to believe that the technical-physical solution of the problem, i.e. the construction of transportation facilities, comes first, before the communication process. Thus, engineers are commissioned to design a transportation plan that will be presented to the public by communication specialists at the very end of the design process.

Communicators are only occasionally part of the planning team from the beginning of the project's conception. Most of the time, they are hired when the plan is ready and it must be presented to the public. And, even in the few cases in which they are engaged from the very start, they never do play an influential role.[5]

Planner Training

Because the commissioners of mobility plans are reluctant to use competent professionals from fields outside traditional engineering, there is little interest, and even less opportunity for economists, sociologists, communicators, traffic psychologists, and so on, to develop, improve, and (mainly) apply new skills in the field. Most of them are all too happy to receive some minor and low-key tasks. I wonder if they ignore the possibility of professional growth, proving their incompetence and lack of self-esteem; or if they are afraid to lose the little space they now enjoy. Another possible explanation could be a mix of the two, of course. It is difficult to escape this vicious circle which we can briefly sketch as follows:

- engineers are overspecialized;
- administrators keep hiring them to repeat their usual approaches;
- thus, engineers do not have any interest in modifying their almost purely technical approach;
- social scientists, generally, are not prepared to assume a lead in urban mobility plans;
- administrators do not even know how they might participate seriously;
- consequently, administrators do not employ them;
- therefore, social scientists and public policy experts have little interest in developing new skills that would fit in with an innovative traffic policy.

In Europe the position of social scientists in planning is even more marginal than in the US.[6] It should be the task of the Universities to break into the technical and reductionist approach and propose new ideas, approaches, and consequent skills.

As mentioned above, administrators usually hire communication and public policy experts at the end of the design process, namely after the final plan is delivered by the engineers. This occurs because the decision-makers suddenly become aware that the engineers' plan – which they had thought perfect – encounters wild opposition from their constituency. Opposition usually occurs when citizens receive a document that proposes the construction of roads and infrastructures that will affect

[5] In Chap. 7, I will describe the case of the sociologist hired by the Northern Virginia Transportation Authority. The leading (engineers) planners hired him merely to confirm their own ideas and to give citizens a superficial impression of concern about public opinion.

[6] See the critical essay by Michel Crozier, *La crise de l'intelligence* (1995) in which the author – a Frenchman himself – mocks the otherwise celebrated French bureaucracy.

their habits, and even their estate values. The political opposition, of course, supports the protest demanding more citizen participation.

Citizens often oppose plans simply because they feel they were not involved in the decision-making process. They sense that the money could have been spent differently. My extensive experience as both an administrator and a planning consultant allows me to be sufficiently aware that it is not that easy to spend public money with the flexibility that most citizens would expect. Nonetheless, the same experience leads me to the conviction that something can be done to facilitate new choices, including more negotiations, and more information to and from citizens, in order to achieve a more legitimate consensus. All of this can become true if we break the monopoly of the engineering approach and entrust the planning process to experts capable of making citizen communication a central issue.

Citizen Opposition and a Chance for Change

Citizens oppose administration plans when they become aware that they are not involved in decision-making and when they realize that the huge amount of money spent in traffic infrastructure building could be used differently. It is difficult to redirect already allocated funds to alternative uses, but the citizens' perception is one of injustice and lack of participation. In politics feelings often count more than facts. Or better; they *are* facts and "facts are (not) stupid things"[7]! The opposition encourages citizens' protests to win consensus, but when they are eventually able to gain power, they do not substantially change the same customary approach.[8] To be realistic, we must admit that there is little possibility to finance a different approach and different public works immediately. Nonetheless, in the mid-long term something can be modified if governments adopted a planning method less inspired by the technical culture and by engineering companies. In this case we could trigger the following sequence:

- infrastructures' and public works' proposals are organized in a general vision;
- thus it is possible to raise openly the political issue of local finance and repartition of resources in alternative projects;
- new projects are proposed which might be financed and implemented in the future, even though it is unlikely they are undertaken immediately;
- new research and new professional skills will develop in bureaucracy and in universities as a by-product of designing new projects;
- last, but probably most important, a democratic process is activated that might overcome the rigidity of the corporatist system.

[7]I recall here an (in)famous quotation by Ronald Reagan who – at the 1988 Republican Convention – emphatically proclaimed that "facts are stupid things".

[8]This does not happen because of the malevolence of politicians who deliberately want to cheat their constituencies. I'll try to explain later in detail how it becomes true.

A planning approach, like the one I am proposing, bears its own risks. For example, we run the risk of substituting competent professionals – although exaggeratedly specialized – with people who are not yet well trained and experienced in traffic planning. It is a risk that must be taken for the sake of progress. It is not even a big risk for two reasons. First, change is not likely to intervene overnight in altering public expenditure. To change laws and approach takes a long time, which can be used to train new professionals with innovative skills. Second, the traffic situation is so deteriorated that it is difficult to make it worse. What is really necessary is to begin thinking differently! "Every long journey begins with one small step" ... but you better know where you are heading to ... otherwise no wind will be favorable.

How to Err in Communication

Sometimes, the political and technical commissioners for mobility and transportation plans hire communication experts. This usually happens only when the plan design has already been completed. They assume that civil engineers know what is to be done. On their behalf, engineers are convinced that they do not need other professionals to help in the design process. Consequently, they know exactly what they need to communicate to the citizens when they have finished their job. Most frequently the possibility to hire communicators[9] is not even considered at the beginning of the planning process and it rather becomes a necessity when citizens revolt against the plan. The leaders in charge of the transportation plan hope that skilled spokespeople can help remove the widespread discontent for the plan they have presented. Communicators are just considered an addition to the process: they are required if something goes wrong, which, by the way, happens all the time.

Planning leaders ask the public relation advisors to convince the citizens that the infrastructures that they – following the suggestion of the sole engineers – have decided to build are benefits to the community. They use the communicators as public opinion manipulators, rather than as real specialists who could create a more open and comprehensible dialogue. We should not consider this move as a malicious effort to cheat the people deliberately. In some cases it might be malevolent, of course, but in general, decision-makers and leaders honestly believe that their plan is what citizens need. There is an apparent disproportion between, on one hand, the engineers' power and the technical culture, and, on the other, the shy and marginalized alternative approach I am proposing. The business-as-usual representatives believe that the citizens do not accept the plan as it is just because they "do not know, do not understand"; hence, decision-makers consider it their duty to convince citizens to like the plan in their own interest. This attitude reminds me of a celebrated quotation by

[9]In this case we should more precisely speak of "public relation" specialists, if not real "spin doctors".

Thoreau: "If I knew for a certainty that a man was coming to my house with the conscious design of doing me good, I should run for my life."

Every social scientist would understand how this customary approach is the best way to create more dissatisfaction, protest, and eventually, a considerable delay in the plan implementation. By the way, it would be accurate to account this delay among the project costs. Unfortunately, engineers in charge of the planning process are not trained to be aware of it. Therefore they do not expect people's discontent, although it regularly occurs, because it is something alien to their way of thinking. Most of the time, the transportation planning agencies do not specifically allocate enough funds for public relations and effective communication at the beginning of the planning process. Eventually, they find some resource in some special fund, included in the "AOB" list, when the situation seems to be out of control. An accident rather than a foresight!

The True Tasks of Communicators in Mobility Plans

"Communication" is a word that derives from Latin. It means "make many into one" or "put together." It implies a multi-party interactive process. We should not confuse communication with "information." This word also derives from Latin, and means to "shape," or "put in form." In this case, there is a subject who "shapes" others, namely the objects to be in-form-ed. The difference between information and communication is that, in the latter, the total amount of knowledge available is supposed to grow as it is exchanged and is made available to different parties: a good word, although somehow *cliché*, to express this concept is "cross-fertilization". In the case of an information exchange, it is merely the transfer of what is already known: as they say in Roman law "*Nemo plus juris ad alienum transferre potest, quam ipse habet*", which translates as "None can transfer to another more entitlement than they themselves hold".

The communicators' task would consist of a series of complex activities aimed at diffusing and linking existing knowledge within the community. This is exactly the opposite of an information campaign which aims at disparaging the opinions that challenge the revealed truth of the plan. This is why communication experts should be part of the planning process from the beginning. I assume someone will now claim that this seldom happens. My point is that even when this does occur – and I agree that it sometimes does, at least in North America – we should nonetheless: (a) carefully examine how the participation of communicators is located in the planning process; (b) argue over the power they are given in the planning team and in the planning process. If we were to foresee the presence of communicators with a leading role in the planning teams, it would be possible: (a) to evaluate, even before beginning to design the physical plan, what the citizens – who might be defined as the plan's co-authors; subjects rather than objects of the planning process – really desire; (b) to favor the circulation of knowledge and the proposals describing the diverse options; (c) to help build a system of interaction among the parties.

It is more than obvious that leaders who retain control of the planning process and have a strong influence on "realistic" solutions (i.e. plain construction) would be suspicious about enlargement of the options offered by citizens as well as by involvement of public relation specialists. These monopolists sustain the "let's be realistic" ideology which, all too realistically, identifies with the solution/project they have had in mind from before the project's conception. The hidden invincible bias lying beneath this idea is that continuous growth in motorization and mobility, no matter how expensive and risky for the environment, would encourage economic development. Of course, this is a position that many people no longer accept and now fight against (see Ray and Anderson 2000).[10] The reaction is that, instead of plans, nowadays we just create operative projects before any open political choice. The reaction to the growing discontent of people concerning traffic plans is to substitute them with projects based on the "let's be realistic" approach. Unfortunately this approach is also endorsed by the citizens who would like to see their traffic problems resolved immediately, and demand incremental solutions that are anything but realistic.

In a more open planning process, planners and communicators could find a role for diverse professionals, such as economists who can use, for example, fiscal tools to favor a retail distribution aimed at traffic reduction. Similarly, psychologists and sociologists could investigate behaviors and social change. Experts in logistics and organization could play a role in thinking about traffic. Unionists and housing analysts could formulate policies that would bring jobs closer to dwellings or vice versa. How much money has been invested in investigating how a housing policy can reduce mobility and the consequent social costs? Architects could design urban environments in which walking would be more preferable than driving, and telecommunication engineers could propose tools which would be able to reduce traffic further. However, all this would still be useless if it is merely conceived as part of a plan designed in the basic civil engineering and transportation framework. All the specialized professionals mentioned should instead begin thinking about traffic with their own discipline's approaches and tools. They should imagine that they were put in charge of designing the whole plan from scratch, without considering themselves as part of an existing project designed by others. And they should refer to their knowledge, their approach, and their mindset. All skills and all professionals should not be ranked according to an encoded hierarchy.

Someone might argue that hiring these professionals and employing their research requires substantial investments and a long time to become operational. Eventually, there is no warranty that it will not result in a complete failure. Failure is possible, of course, but it would be exactly the same failure and the same time wasted by the "let's be realistic" ones! These realistic, self-appointed practical people are nonetheless allowed to go on repeating the same mistakes without raising the same perplexities. How much time has been employed to decide how to build beltways and underground rails that have never definitely solved any real transportation problem and have, more often than not, worsened it?

[10]This attitude is much more common in affluent Western countries than in growing economies.

We have been failing in traffic policies for decades because we are no longer asking in a systematic, scholarly and professional manner the following essential questions: (a) why and how much do we need to move? (b) does an overall increase in mobility automatically imply a progress toward some socially shared goals? (c) what are the interpretations of the context – social and geographical – on which we operate? I assume that, if we only attempted to give serious answers to these questions, we could envisage different and more effective urban mobility policies, at the same time realizing that consensus is part of effectiveness.

Relevant Questions vs. Irrelevant Data

At a recent Conference, I heard an enthusiastic young scholar, who had won awards for an article of hers published in a major journal, saying: "I found wonderful data, so I decided to do the research!" The research consisted in an econometric elaboration of those data. I wonder if we do research just because we have data or because we want to find something useful in order to do something in which we believe and, in view of that, we look for possible numerical data and other information. Thus, if there is no data available, we should either collect them or adopt different research tools. We all live in the mediocrity of the world and know that some research is done just for the sake of academic careers. If we want to be even more optimistic, we can add that, using existing data allows social scientists to elaborate models and analytical tools that might be later applied to other situations. Hence, we should not criticize the young scholar's declaration too severely.[11] However, as social scientists, we should at least be ashamed when caught in such a *faux pas*. Too many researchers focus on improving analytical models using the existing data and much less look critically for new meaningful data/information that is lacking.

Traffic congestion is not the cause of the problem, but merely the result of failing to solve other problems. Citizens travel for a number of reasons that are not only geographical and economical, but also social, psychological, and anthropological. This is why changing mobility patterns may be difficult and even traumatic for many. It is also a good reason not to enforce a single mobility policy on all citizens as if they all had the same preferences and needs. We should be able to offer alternatives to the many who do not feel comfortable with the current mobility patterns.

We do not move only because of economic convenience or because we need physical access to goods and services, but for a variety of complex motivations, not necessarily rational. From a geographical point of view, we travel to gain access to a service that is not available everywhere. From a utilitarian point of view, people move because

[11] I am not complaining that such research was carried out by the young scholar. I do criticize, instead, that this kind of research is encouraged and awarded credit and prizes because it is at the center of academic research and concerns. As a matter of fact, it is easier to award a prize to a falsifiable quantitative research than to risk long discussion about some more debatable argument.

they make a higher profit by being in another place. Merchandises are transported to places where they can be sold at a higher price having a higher utility. The goal is to obtain an advantage by being somewhere else. However, travel generates costs in terms of energy, infrastructures, time employed, transportation means, operation and construction, pollution produced (air, noise, waste), congestion, etc. We generally do not calculate some (most) of these costs in the decision-making cost-benefit analyses simply because we are accustomed to overlooking them.

Let us for the moment limit our analysis to economic convenience and comfortable access to services. We still have two alternatives. We can either: (a) improve the accessibility of services by making them available to a wider area, while increasing mobility; or (b) enhance a diffused retail distribution of services in the neighborhoods. As everybody knows, mobility plans completely neglect – with a few exceptions which we will soon describe – the second hypothesis, which is alien to the training of most professionals who design plans. Considering the dreadful situation of traffic in cities, it would be wise at least to examine the second hypothesis and carry on the necessary research to find solutions. Unfortunately, in colleges and research institutions, most researchers process only data that already exist and prefer not to identify new approaches calling for innovative data collection. This would be the so-called "independent research" for which they are supposed to be paid.[12]

If we assume that, in order to supply necessary services in urban areas, we should increase accessibility through more mobility, we are able to advance two strong, implicit, and conservative hypotheses: (a) a widespread retail distribution is unlikely to be established; (b) public policy ought to focus on building the necessary infrastructures to allow existing services to be available to most people. This is what we have been doing for decades without seriously questioning the citizen's level of satisfaction. On the contrary, if we assume that we should increase the availability of services, including housing and jobs located at walking distance, the government could then shift public funding from one sector to another. This could occur by shifting money from road construction to retail distribution, information, housing, schools, medical care, etc. One should not consider the two options mutually exclusive. In order to prepare for a change, it would be enough just to present an elaborate alternative. Once presented, investors could foresee a potential market and it is likely they would propose new projects.

Expectations and the 10% Plan

Just a few weeks before the end of his third mandate, the Italian Veneto Region Governor, Mr. Giancarlo Galan, decided not to run for his fourth mandate. At the last Council session he tried to win the vote for construction of three new waste incinerators.

[12] At present, regrettably, social research and academicians' salaries are also to a large extent funded by non–academic institutions and this cooperates to inhibit change.

The act did not pass because of the usual NIMBY syndrome that often occurs in these situations that are even more likely before elections. The Governor's act did not pass also because, during his 15 year mandate, he never approved a waste disposal and treatment plan that took into consideration the possibility to enact a policy aimed at reducing waste production, but rather presented the incinerators as the single and ultimate solution. Somebody suspected that there was no interest in reducing the production of waste because this would have made more incinerators unnecessary; these facilities are notoriously owned by big business, funded and subsidized by government money. If the plan had been approved it would have included some more general concern about how to deal with the overall waste treatment problem. If a plan that included a waste reduction policy had been proposed, part of the opposition might have accepted a compromise, i.e. one or two incinerators to face the emergency, but no more waste growth or, even worse, no import of waste to burn for money from other areas thanks to an exceeding capacity of the incinerators built. The incinerator's constructors and the Governor preferred the very arrogant strategy to avoid any planning and to push instead for having more waste and more incinerators because they were confident they would win council approval. In this case the Governor's strategy failed, but the consequence is that now there is neither a waste reduction plan, nor incinerators and there are not even grounds for a compromise.

The waste treatment problem is similar to the traffic problem. The strategy I propose is not expected to work on the whole mobility system, but just on a portion of it, perhaps 10% that can realistically and immediately be managed. With this 10% quota, we can likely intervene to control mobility inflation. Doing so, we can change expectations, which are a major factor in traffic and economic behavior and are now featured to generate an untamed growth. If we expect that the solution to our mobility problem will be the construction of a new road, we are likely to buy a new car and a new and less expensive house, no matter how far it is from our job. Knowing this, the business community will invest in road and car/bus/rail construction and in scattered land use developments, such as shopping centers that are absurdly located where jackrabbits live rather than where people live, as Gottlieb wittily argued (2007). On the other hand, if we expect that the solution to our mobility problem will be more telecommuting, more subsidized loans for housing in high-density areas, fewer taxes for neighborhood retail distribution, and so on, we will make our decisions and organize our lives accordingly. From a psychological point of view it is proven that if we expect that a traffic jam problem will soon be solved by the simple construction of a new road, the present discomfort will become more acceptable and protest less likely.[13] Even if the actual construction of the new road is delayed for years and sometimes for decades, this entrenched expectation hinders thinking about possible alternatives.

[13]This sentence and my previous arguing about the role of ideology is inspired by several essays by Albert Hirschman with whom I spent a short period at the Institute for Advanced Study at Princeton (NJ) and whose essays I eagerly read since I was an undergraduate. Here I specifically refer to *Exit, Voice and Loyalty* (Hirschman 1970) and *Journeys Toward Progress* (1963).

This alternative approach would only be utopian if we were talking about targeting the mobility problem immediately and in its entirety, but it is all too plausible if we deal with only 10% of the mobility problem. In general, we approach the mobility problem presuming that we still live in a mass society, namely, we envision a policy which would satisfy the average citizen. Most planners still think they should (and can) operate on all city mobility patterns. This assumption would be acceptable if the citizens' attitudes and preferences were relatively uniform and shared. On the contrary, in contemporary western affluent society, citizens groups – the urban tribes, as the British anthropologist Desmond Morris (1967–1969) labeled them – are very different from each other.[14] For this very reason, we should speak of urban "mobili*ties*" plans rather than of one standard mobility plan. There are many alternative traffic policies that could respond to the diverse needs of a diverse citizenry. The first goal should be to multiply the number of policies from which people can choose.

We must admit that this approach is risky for the most conservative among the business community, those who operate in traditional mobility-related facilities. Adopting this approach to transportation planning may well (and is intended to) prepare a sequence which affects progressively larger and larger quotas of the overall mobility system. In this way, we create the possibility of matching the needs and the preferences of a greater number of citizens who are now hindered from expressing their preferences and fulfilling their goals. Consequently, this approach can become a political platform on which to aggregate consensus and constituencies.

Someone could reasonably argue that, in the end, different solutions to traffic problems are simply not possible. The problem is that little comparative research has been completed and only small amounts of public funds have been allocated to search for innovative and alternative projects. When undertaking research, one should be aware that there is a risk of finding nothing useful. But if we do not do any research, this risk becomes a certainty!

The most practical part of any traffic policy is to understand why people need mobility, not how much they now travel. There are many reasons why people need mobility: some are obvious, others are more uncertain, hidden and difficult to gauge. The fact that it is difficult to identify the more profound and possibly meaningful reasons why people travel does not imply that these reasons do not subsist, not that they cannot be identified. It is often a typical psychological reaction to react to a problem by denying its existence. Don Ferrante, a character in the famous nineteenth century Italian novel "*The Betrothed*" by Alessandro Manzoni, was unable to give an Aristotelian definition of Plague (neither substance, nor accident). Consequently, he denied contamination, hence did not take any precautions, and eventually he died of it, however convinced he was dying of something else. We are not that different from Don Ferrante when we deal with traffic and environment.

---

[14]Regarding more recent and more socio-political descriptions of the transformation from a mass society to other forms of social structure, I refer to Michael Hardt and Antonio Negri, *Multitude: War and Democracy in the age of Empire* (2004), and Zygmunt Bauman, *Postmodern Ethics* (1993) and his following publications.

The Plan Preparatory Studies: Conservative and Innovative Approaches

At this point two questions ought to be raised: (a) Who should be in charge of new studies and researches? (b) In particular, who should be in charge of raising new questions directed toward formulating innovative solutions to decade old problems?

Scholars and intellectuals should be the source of new ways of thinking and operating. Unfortunately, they rarely function in this role. Administrators generally are not in a position to pose innovative questions to broadly based open-minded researchers, but rather must put their trust in the hands of specialists and experienced professionals. Researchers, many of whom live in a protected academic milieu, tend on one hand to explore theories for which there may be no practical applications, and on the other hand to supplement their academic income by undertaking low-level, conventional professional studies that lack innovation. Between these two extremes there is the possibility of applying some good theory (i.e. critical thinking) capable in the long run of changing the world. Because this rarely happens, we should not be surprised that conventional analyses and surveys prevail in evaluating real policy proposals. The most enlightened administrators should reach out to Universities – which to a large extent, in most European countries, are paid with public money – to deal in a visionary manner with middle-term planning. Regrettably, such planning is currently seen as a waste of time and few succeed in keeping the right balance between short and long term policies.

If we think we need to solve the problem of service availability by increasing accessibility to existing services, we are making two strong and conservative implicit political hypotheses: (a) the availability of services in an area cannot be increased; (b) consequently, public expenditure must favor accessibility by building transport infrastructures linking the existing services. Any evaluation of the users' satisfaction and of the environmental and economic consequences is not taken into consideration. If, on the contrary we think that we should increase the availability of services in an area – including jobs and dwellings – it will be necessary to shift resources from one productive sector to another, e.g. from construction and management of transport systems to organizations of retail distributors. The two options are not mutually exclusive. In order to prepare for change, it should be enough just to forecast change. Once new options are presented, they become realizable, and market operators begin to design projects suitable to the new possible situation. Sadly, everyday experience teaches that this way of thinking, which is based on a sound critical logic, is treated with contempt by many who have an interest in the status quo. Regarding the previous point (a), there are alternative solutions available. The concentration of services is indirectly financed by the structure of public investments, and it includes also schools, hospitals, shopping centers, as we will see in the following sections. The construction of a shopping center implies an obvious traffic growth. The justification is often that concentrated large shopping centers have a positive effect in reducing retail prices. When faced with objections about the possible traffic that the shopping centers will produce, the answer is given that the developers will pay for the infrastructures needed and they will build parking space and roads. The offer might appear adequate initially, but it does not take into consideration the traffic produced throughout the entire urban system. To better understand this process, we need to consider that nowadays retail distribution companies are huge enough to directly and heavily affect the financial markets as much as construction companies do. Often retail distribution operators and construction companies are intertwined in the financial market. Hence they are in a position to delineate long term strategies based on a vision that they have the power to sustain. Unfortunately, this vision can only be conservative being elaborated by the company's bureaucracies. The same long term vision is lacking also in the political milieu, and we cannot hope for an enlightened innovator: the third millennium capitalists can be creative on the financial markets, but they have lost control of production that is now dominated by bureaucrats.

To initiate a planning process based on an alternative public expenditure, we would need to compare the long term costs that a public administration sustains in order to realize public works more or less directly related to transportation, with the possible public expenditure aimed at putting residences nearer to working places, in offering services dispersed in the area, and in general, improving the quality of life in each neighborhood. If this research is promoted, by itself it would open new visions about the future of cities and would create a market for alternative projects. At the same time, a study about the deepest motivations of physical mobility would prepare people to think in new ways and to reconsider their own behavior.

Chapter 6
Urban Space and Mobility Policies in Europe and in North America

> *.... and I find some shade in a park and try to rest. It isn't restful....*
> *The streets of this town are broad, much broader than they need be, and there is a pallor of dust in the air ...*
> *Everything is more run-down and mechanical-looking, and sort of randomly located. Gradually I see what it is.*
> *Nobody is concerned anymore about tidily conserving space.*
> *The land isn't valuable anymore. We are in a Western town*
>
> Robert Pirsig, "Zen and the Art of Motorcycle Maintenance"

Abstract European urban regions are becoming more similar to American low-density metropolitan areas in the organization of space and in people mobility. On the other hand, in North American cities, high-density neighborhoods and renewed European-like downtowns have become popular. Everywhere in the Western world, it is difficult for any politician to refuse further development. Therefore, although the most innovative planners in the most environment-concerned cities focus on issues such as neighborhood life and go for mixed-use zones, they are not quite successful in limiting urban and traffic growth. Contemporary society is actually formed by several social groups with preferences, experiences, and values often radically diverse. Thus, we cannot offer one single option for urban mobility to respond to a variegated demand. Nonetheless, the mass-society/mass-transportation paradigms still dominate planning and transportation policies in most urban areas in the world. Because of the major public works funding system, it is difficult to avoid the approval of mobility infrastructure, no matter how detrimental, helpful or useless they are. The public finance problem of how to pay for mobility infrastructure is related to: (a) the production system and technologies; (b) basic political values; (c) administrative geography. According to a common planning approach different scenarios for future city and traffic developments are designed. Then, one of them is selected. Another solution is possible: instead of selecting only

C. Poli, *Mobility and Environment: Humanists versus Engineers in Urban Policy and Professional Education*, DOI 10.1007/978-94-007-1220-1_6,
© Springer Science+Business Media B.V. 2011

one of the scenarios, we should respond to citizens' diverse needs. This approach responds to the opportunity to favor creativeness in administration.

Keywords Europe • North America • Urban geography • Pacific Rim • Mass-society vs. fragmented social demands • Area and administration • Infrastructure funding • Ethics and politics • Constructors' lobbies • Taxation • Government (federal, state, local) • Urban secession • Metropolitan areas • Technology • Cultural creatives • Creative class

Urban Geography in North America and Europe

Getting Closer or Moving Apart?

Until 30 years ago European travelers visiting North America were deeply impressed by the different organization of settlements and urban space. Today, on both sides of the Atlantic, urban settlement patterns are still different, but European urban regions are becoming more similar to American low-density metropolitan areas in the organization of space and in people mobility. In Western Europe, despite a widespread grassroots opposition, large shopping centers have proliferated and become a considerable portion of the retail distribution system. This is even more prominent in developing Eastern European countries. Adopting a zoning approach – the very same idea they have, in principle, been opposing for a long time – European planners have designed suburban, low-density residential areas, catering to the interests of developers. As a result, urban plans have encouraged the use of cars and mass transit. On their behalf, Americans are inspired by the traditional European urban settlement pattern. In North American cities, high-density neighborhoods and renewed European-like downtowns have become popular. Even several new shopping malls have been designed according to some fake – and frankly tacky – European downtown medieval style. A growing number of American singles, professionals, and small families find it particularly attractive living in areas that resemble the typical historical centers of European cities. Upper middle-class small families and singles, who represent a growing quota of the population, are now in the condition to satisfy this demand of theirs in new or renewed neighborhoods. Let us note that this return to the downtown is the opposite of the suburban American dream of the fifties. In the future, this back-to-downtown fashion, might create a larger market and become available to larger sections of the population.[1] The growth in gas prices, the change in family structure, new technology, and behavioral trends – sometimes encouraged

[1]According to what I claimed in Chap. 2, we may sustain this trend by enacting "zero growth" municipal laws which forbid any further land development. This would be the seed for an environmentalist "ideology", or at least a very strong principle to start with. In some European municipalities, this movement is shyly beginning.

and made fashionable, e.g., by TV series such as *Sex in the City* and *Friends* – might strengthen and hasten this evolution.[2]

The old technologies and the social organization made available to the American affluent mass society of the 1950s, are proving convenient also for contemporary middle-low income Europeans. Many Europeans now run away from the expensive city centers and from decaying peripheries as the American lower middle classes did in the fifties and sixties. They move into new suburban areas where housing is cheaper, and roads and public transportation are heavily subsidized by governments.[3] We can find another similarity between Europe and the US that is not often noticed because of persistent biases regarding public transportation: in Europe, people use cars as often as in the US, and are, consequently, demanding more roads and highways. Certainly, there are still differences due to urban geography and mass transit organization, but they are becoming less and less relevant, the trend being a constant decrease, both in the share and in absolute terms, of persons using collective transportation. It is true that European commuters still drive fewer miles than Americans do, but also this gap is rapidly closing as we build new beltways, shopping centers and new suburban office developments.

The Pacific Rim

In 1995, during a semester at the Queensland University of Technology in Brisbane, I studied urban policies of the so-called Pacific Rim cities – namely Brisbane, Vancouver, Seattle, and Portland. I learned about innovative environmentalist planning approaches, originating in the American Northwest, the land that is so far-west that it could be equally well the farthest-east. The Northwest is populated by Asians and Native Americans whose culture is now appreciated by descendants of the old European colonists who conscientiously tried to exterminate them until a century ago. In 1995, environmental considerations were probably much more emotional than nowadays: they included New Age ideas, Eastern religion environ mental ethics, and an elite radical approach. The communities of Portland, Seattle, and Vancouver introduced grassroots democratic and environmentalist ideas into the public debate. Local politicians were implementing more liberal and environmentalist policies and green leaders, who began their career as local activists, gained influential positions in the government.

Thus, in the last decade, thanks to that environmentalist, multicultural and grassroots atmosphere, these municipalities were able to adopt innovative methods and tools suitable to address traffic problems more effectively than the typical heavy

[2]It is not a thorough paradox to hypothesize that a successful TV movie, which advertizes an environment friendly behavior, would reduce pollution more than some sophisticated International agreement on emissions.

[3]Note how recently, both in the US and in Europe, car industry has been straightforwardly subsidized to cope with the 2008 financial crisis.

road construction. When I met with some American Pacific Rim city planners, scholars and political activists, initially I felt as if I was preaching to the choir. Later, while talking about the political framework of traffic policies and the new tools to contain traffic growth, I realized the choir still had a few reasons to listen to me. The major goal of Portland's and Vancouver's radical and influential urban thinkers was the reduction of car traffic. This would result in a widespread promotion of biking and the use of public transportation which was, in fact, no longer really radical. The North Western radical urban thinkers also proposed more densely populated areas. This also was not completely original, but in the case of Vancouver, Portland and Seattle, real policies for densely populated areas have been seriously implemented more than in other cities where they were just spoken about. The economy of these cities is booming and their population is rapidly growing. They are successful thanks to the quality of life they offer which is inspired by a certain environmental consciousness. Moreover, these cities have become the hub for many youth attracted by the trendiest music and arts. As a consequence, the "tragedy of the commons" syndrome (Hardin 1968) is occurring: excessive economic and demographic growth is an environmental dis-value *per se* as it implies unavoidable higher impacts, which, in the case of transportation, means the need to move more people on a daily basis, in absolute terms. In this respect the difference from Europe is remarkable. Economic development of middle-sized European cities is not associated so much with demographic growth. Obviously, successful cities in Europe may also show some population growth, but the quality of life that those cities offer does not attract a significant and immediate immigration as it happens in North America where population is more inclined to move.

In the United States, as in any other place, it is difficult for any politician or ruling class to refuse further development. Land values are officially calculated only in monetary terms. Developers and political decision-makers are not requested to assess possible considerations about other values incorporated in undeveloped land such as precaution value, intrinsic value and existence value, to name but a few that have been broadly analyzed in environmental ethics literature.[4] In environmental ethics and policy these values should be preserved beyond the economic short term customary use-value. From an ethical point of view, however, the question becomes more complicated for politicians. When they act as political leaders, decision-makers ought to present their beliefs to their constituency. On the other hand, when they act as representatives of the people, they should implement the mandate received by voters, and do so in the best way possible. If individual citizens – not the vested interests – still want development, then that development cannot be refused. Possibly it can be implemented in the best viable way. Democracy is often boring or mediocre, but there is nothing better we can do. For this reason it is only in smaller cities that it will be possible to implement real and radical changes. In smaller cities people may affirm a widespread conviction about an alternative idea for mobility and, in general, life-style. There are interesting examples of such towns in North America and in Europe whose achievements should be advertised. In my

[4] See above Chap. 2, note 4.

recent essay on "*Flexible Cities. A Revolution in Urban Government*" (Poli 2009), I have extensively theorized the potential revolutionary change that might take place in smaller cities.

Even the destiny of progressive cities such as Vancouver, Seattle and Portland, ruled by enlightened leaders, seems to be toward further growth. Although scholar and former politician Gordon Price[5] from Vancouver admits that he will try to oppose the construction of new roads and bridges as well as to counteract politically what in this book has been defined as traffic inflation, there is little possibility he will succeed. In the best case, mass transit facilities will substitute for a few road projects, but overall population and mobility growth seems impossible to curb. More or less explicitly, and without any serious questioning, most, if not all, of the people in charge of mobility, even in these environmentalist cities, maintain that the building of more public transportation infrastructures will automatically pull people out of their cars, and therefore solve the congestion problem. Of course, this is a matter of faith rather than rational arguing. In these cities there is at least more awareness about the need for operating with tools other than mass transit. But awareness is still a step or two behind action. The sustainable development approach, which favored a shift toward more environment-friendly technologies, is working against more extensive changes in transportation policies, as I have repeatedly argued.

The most innovative planners in the most environment-concerned cities focus on issues such as neighborhood life, citizens' participation in urban design, dwelling and working in the same area, i.e., they go for mixed use zones, retail policies, and so on, rather than zoning and new huge shopping malls. I list some of their projects and proposals for reducing traffic in Chaps. 12 and 13. Some of them have been adopted, others just considered in a number of US and European municipalities. Most of the planners I met during my lecturing in the US seemed interested in some of the ideas I presented which I drew from the European experience or autonomously elaborated. In cities where a long tradition of environmental concern had been established, I realized that my proposals, though radical, were understood and accepted by civil servants and politicians because the ground had been prepared for decades. Maybe they will be implemented in the near future when the time will be ripe. It will not be easy to oppose the general trend because other transportation plans I have examined both in Europe and in the US are far away from the beliefs I share with the most innovative city planners, especially the ones I met in the so-called Pacific Rim.[6]

Everywhere in western affluent and pluralist societies, it commonly happens that most innovative ideas for traffic reduction are actually adopted, financed and

[5] I interviewed him in May 2007 in Vancouver.

[6] One of the first authors who began thinking about reducing traffic is pointedly David Engwicht (1989) from Brisbane. In the following years up to now, Engwicht has been active in promoting and thinking about many projects intending to reduce traffic. His designing activity has been quite influential among environmentalists and his ideas have been applied in several small cities. Engwicht's work has been political as far as he opened a new way of thinking about traffic, but he did not go beyond the simple proposal of piecemeal, though often witty, solutions. Thus, he fails to elaborate a political thinking about traffic. As a matter of fact, that was not even his goal, while it is the main goal of this essay.

implemented. There are hundreds of smart projects and plans that would be efficient if only they were adopted as substitutes of the traditional approach which relies on road and mass transit construction. Unfortunately, as they merely represent an addition to the usual traffic growth oriented policies, they cooperate paradoxically in supporting the mobility growth trend. We can welcome projects such as renting bikes, car-sharing and car-pool systems, and so on, since we consider them as educational, but they are little more than advertisements for real environment-friendly solutions to mobility. They prove that alternatives to driving cars can be proposed, but as long as they are not substitutes for mobility rather than for other transportation means, they are ineffective.

Contemporary Western Society

Beyond this geographical similarity, we argue that both in the US and in Europe, contemporary society is somewhat different from the one we were used to analyzing some decades ago. When planning for large infrastructures, such as traffic and transportation, we should consider that the so-called mass society does not exist anymore. Contemporary society is actually formed by several social groups with preferences, experiences, and values often radically diverse. We do not refer just to the intense immigration that now makes Europe more similar to the US from the point of view of the ethnic composition of the population: the diversity also involves native groups and is almost independent from the most common sociological classifications which are based on traditional indicators such as social class, income, and ethnicity. We do not want to deny that traditional indicators are still important for developing some public policy projects. However, we must add many more new indicators that are growing in importance compared to the ones usually adopted.

If these additions are accepted, then urban research should be directed to identifying the needs and the demands of diverse groups of citizens. Their preferences are not only determined by income levels, but to a much larger extent by life-styles. This is a general consideration that applies to several fields of public policy and of course includes transportation and mobility-related infrastructures. It seems obvious that we cannot offer just one option for urban mobility if we want to respond to a variegated demand. Nonetheless, the mass-society/mass-transportation paradigms still dominate the planning and transportation policies in most urban areas in the world. This implies a specific funding model.

Infrastructure Funding and Government Size

Local and Central Finance: An Ethical and Political Perspective

As I reported in Chap. 5, citizens are often asked to accept a plan they do not know much about until it is almost complete. Most of the constructions listed in the plan/project are the ones that the decision-maker/politician has suggested because it was possible to finance them or part of them (but nothing else) with federal and/or state funds. Decision-makers in local governments compete with each other to get those State or Federal funds. The lobbies of the construction corporations had previously operated at federal and/or state level to have those specific works financed. Because these infrastructures are significant, there are very few firms that can realistically apply for the job. In most cases, as typically happens in oligopolistic markets, they tacitly agree and share out the possible contracts in advance. In Europe all companies are officially allowed to compete on any national market, but national and regional governments still try to somehow favor their own companies. Although the situation is changing, so that a European market for public works is now in function, an oligopolistic market is well established.

Having Federal and State funds available for public works is a great asset for local politicians who can: (a) use their power by spending money that they do not collect directly but is transferred to their constituency by the central government: (b) prove to tax-payers that they are able to return that money into the community. The problem is that this Federal and State funding can be used only according to what is decided outside the community. Apparently, there is no comparison between the power of the developers' lobbies and the citizens' influence of small constituencies at the State/Federal level. Local politicians are put in front of the difficult dilemma if is it worthier and more ethical: (a) to renounce some possible Federal or State funding rigidly assigned to build something that is not considered essential for the community's well-being[7]; or (b) getting the constituency money back and using it in the best way possible. The most probable – and yet ethical – decision is the one at point (b), so that the UMP becomes a list of projects that can be managed locally, but are decided to somewhere else, where local representatives and citizens cannot directly exercize their power.

The public finance problem of how to pay for mobility infrastructure is related to: (a) the production system and technologies; (b) basic political values; (c) administrative geography. Regarding point (c), when metropolitan areas expand on territories administered by more and more local governments, it becomes more likely that the infrastructures are decided at State and Federal level. The proper functioning of the metropolis becomes a national concern rather than a local issue. Metropolitan areas expansion is thus encouraged by specific available technologies and by the funding system applied. As in a virtuous/vicious circle, the geographical perception of metropolitan areas is also promoted by the opportunity to build huge infrastructure. Another inconsistency of the mobility infrastructure financing system regards urban concentration. In the last 50 years, we have experienced both people polarization in large urban areas at national level, and suburban sprawl of cities. This is in part a consequence of the technologies that have been applied, and in part it is the effect of an unfettered urban rent. In other words it is not an unchangeable destiny, but the outcome of a series of deliberate political choices. Thanks to the availability of new technologies, we are now in the position to begin thinking

[7]Note that Federal funding is made of local citizens' taxes that are brought back to the community thanks to the political power and ability of the local politicians.

about a different urban form which will re-evaluate small centers and a more even distribution of people on land at any scale. Nonetheless, customary mobility policies still focus on daily mobility – i.e. commuting – instead of thinking of the option of drastically reducing commuting and possibly favor other projects such as telecommunications, occasional weekly-monthly long range transfers among towns, self-sustained communities in terms of jobs, dwellings, recreation, services and the like. As a matter of fact, decisions are not made by Municipalities, even though a significant amount of money is appropriated for projects that have a local impact. It is the Federal or the State who steers the metropolitan infrastructure policy whereas the municipal governments are only competent for planning the way infrastructures are built and not what infrastructures should be chosen. Even worse, most of the time local governments are requested to cooperate with extra funds of their own to complete an infrastructure that is partly funded by the Federal or the State, but completely imposed by them.

There is virtually no possibility for the municipal authorities to escape conditioning by higher level governments. In fact, local administrators have an ethical-political obligation in recovering the money which tax payers had paid to Federal and State, and in using it for the good of their constituency. It would neither be moral nor logical to act differently.[8] Even if the infrastructure funded by the Federal or the State is not considered that necessary by local administrators, a cost-benefit analysis will assess the following: if the municipality does not get the money back for that specific public work (decided elsewhere), the same funding will be completely lost and the municipality will not be allowed to use it for alternative projects.

For example, a Federal agency or a European State ministry assigns funds for building light rails in the country's urban areas with more than 500,000 inhabitants, each of them administered by a number of local governments. The intervention of the higher authority is justified by the fact that many local governments are supposedly interested in that infrastructure, namely the light rail. Local governments in large urban areas are very different from each other in size and population. If a major city could autonomously finance a light-rail regional transportation system, it would be out of question that small local administrations could build it. But the light-rail is supposed to be in the interest of all municipalities, i.e. of the metropolitan area. Things thus standing, the Agency or the Ministry are entitled to intervene with their money.[9] Citizens and local politicians might have originally

[8] Except in the very improbable case that an administrator can prove that the public work financed with Federal funds has such a negative social and environmental impact that is better to refuse the funds.

[9] Their capability to impose or suggest the characteristics of the different projects in competition for public money depends on the relations between the central and local governments. The power of the central states and of federations has progressively grown in the last 50 years in Western countries. Moreover, these characteristics and technologies have often been suggested by lobbies which are interested in selling their technologies to the municipalities selected to receive the money. It is much easier for big companies to lobby at the central level rather than responding to thousands of calls for tenders.

preferred to improve the bus transportation system by buying electric buses operating in the single municipality. This approach would have favored more compact settlements. The light rail, with the standard characteristics necessarily imposed by the Ministry or the Agency, would encourage urban sprawl, instead. Nonetheless, if the administrators renounce competing for Federal or State money, they would lose both the possibility of having the not-so-convenient light rail and running the electric buses in small compounds as well. Of course, they can decide to refuse the money because the public expenditure would create more damage than benefits, but the cost-benefit-risk analysis for the approval of the single project is biased since the beginning, because of the manner in which the money has been collected and redistributed.

For a less biased traffic policy, economists in public finance should study the effects of public expenditure structure on decision-making. It is likely that a shift toward more autonomy of communities in deciding how to use their own money, or the money returned by states or federal, would bring a considerable change also in the characteristics of the solutions given to mobility problems. But economists and public finance specialists are heartily encouraged not to mess with these evaluations nor have scholars been active[10] in proving the inconsistency of this decision-making structure. Some research, of course, has been proposed. A marginalized young scholar, speaking at a huge Conference in a sleepy session, one day might have reported on this problem, possibly providing a sound argument proved with abundance of data. But, missing a general theoretical (ideological) framework in which to locate it in a political perspective, it passed away unobserved even though this kind of research should win the limelight in the traffic debate play.

Central, Metropolitan, and Local Governments: How to Proceed

Because a middle size city administration needs to apply for money returned from the central state or Federal government in order to build (presumed) necessary infrastructures, then creativity – which should be a social value and has been a driving force of early modernity – is *de facto* inhibited. In fact, the central government will impose conditions on how the money is to be spent.[11] These "conditions" will be "suggested" by the few companies able to lobby at national level and to contract huge public investments.

Instead, if we conceive of metropolises as a number of small towns, the perception and the consequent policies may dramatically change. Transportation plans

[10] Most scholars are neither active politically (as they ought to be) nor (consequently) do they show scholarly enthusiasm in studying an alternative fund distribution pattern.

[11] Modernity is revolutionary *per se*. Modernization is a process meant to overthrow old structures and beliefs.

would consider limited areas whose mobility and livability problems ought to be resolved as much as possible from within and/or in circumscribed areas. Only secondarily ought we to aggregate them in broader compounds. We have developed and used technologies in order to extend our control over broader and broader spaces. For many decades, at times when it was not possible to solve an urban problem locally, it was transferred to a higher level of government, without even trying local solutions. The time is ripe to develop technologies suited to improve local quality of life without more physical expansion. The old approach has been failing for at least 30 years. Once passed a certain dimensional (area and/or population) threshold,[12] all the problems have worsened and their solution, which was theoretically supposed easier, has become in fact impossible. Metropolitan Governments, established all over the world to manage sprawling urban areas, have proved ineffective everywhere. Geographers' decades-long attempt to identify urban regional units – which have been in turn defined as Standard Metropolitan Statistical Areas, Standard Metropolitan Area, Metropolitan Labor Market Areas, Urban Fields, and more – has never led to an effective policy.[13] The failure has been both: (a) political, i.e. little or no power has ever been delegated to metropolitan governments either by the Municipalities or by the States; and (b) operative, i.e. the problems have been worsened rather than solved. We do not object when a municipality, after having tried to resolve its mobility problems autonomously, seeks to coordinate with neighbor governments. On the contrary, we sustain the objection when the same municipality aims at solving its own problems by exporting them to a larger area. It would be useful to develop comparative studies about local governments in middle-size cities. Diversity is a powerful generator of creativeness. Comparisons between local governments in Europe and in the US, have not been examined enough, hence more research should be promoted in this field in order to make use of and understand the experiences of numerous smaller cities.

The proposal emerging from these considerations and the general approach is to begin planning from a smaller scale, e.g. 20,000/40,000 inhabitants units. In these

[12] In planning, economic and geographical theory, the problem of dimension and a specific threshold theory occupied many scholars in the 1960s and 1970s. For an early summary of it, see Kozlowski and Hughes (1967). At the time the problem was how to overcome a population or an area dimensional threshold that would make an activity or an infrastructure competitive on the market. The leading idea was to attain "increasing returns of scale" – the larger, the more efficient – and the search for bigger dimension was the main goal. However, also at those times, in some situations it was evident that "decreasing returns of scale" were already operating in some situations.

[13] There are of course some exceptions like in the celebrated case of Barcelona that in the eighties managed a successful development plan. Now, that major achievement belongs to the past and Barcelona has the same administration problems of any other European metropolitan area as Pascual Maragall – the city Mayor for 1982–1997 and then the President of Catalonia from 2003 to 2006 – also admits. The very fact that Barcelona's case has been so admired and studied proves that it was an exceptional case. Many tried to imitate the Catalan capital's strategic planning, but nobody had the same success.

areas we should think about minimizing the need for either mobility and for importing services. Moreover, if we are able to maximize the flow of information among the members of communities, we can aspire to enlarge internal markets without physically expanding them. In other words, I recommend intensifying neighborhood-based community life. When I say that we must "think" in this way, I mean that when administrators propose new infrastructures or any other project, they themselves and/or the citizens should raise the explicit questions: (a) does the project fit with this approach? (b) what are the consequences of it on the enlargement of the metropolitan area? (c) is metropolitan area growth good or not? Once this micro-medium scale planning is designed, then no one will prevent any governing body from agreeing with another in order to coordinate their policies. The goal would then be to coordinate and integrate the movement of people among communities which have tried to become – as much as reasonable – self sufficient. These communities would try to solve their mobility problems by looking inward instead of exporting them on larger and larger areas, where their solutions will clash with other communities who have been carrying on similar policies.

Both in Europe and in the US, most of the technologies applied in transportation policies have been developed to treat the mobility problems of large metropolitan areas. In other words, large metropolitan areas have colonized smaller cities with their technologies. In some cases, middle sized cities have "bought" – or, better, the Government has sold – infrastructures (such as underground trains) that are over-dimensioned regarding the existing demand. In these cases, the problem is not the occasional waste of public money. It is that by doing so, the cities have imported problems that might have been avoided. For example, urban sprawl is encouraged by regional transportation networks (roads, buses, trains). It is equally encouraged by shopping centers that are efficient only if the area from which the customers are drawn is large enough. For the draw of consumers to be large enough, subsidized highways are needed. Then the consequence is that more public money is needed to keep the infrastructures working.

The Best Scenario vs. the Multiple Scenario Approach

It has become a quite common planning approach to design different scenarios for future city and traffic developments, and then to choose one of them, possibly with some integration drawn from the other scenarios. I do not completely reject this method. It is useful in that it allows us to think and present different possibilities, each of them consistent with specific values and goals. The internal coherence of each scenario is an important asset of the plan, but as soon as we choose one, the citizens who made another choice are forced to accept a scenario which does not respond to their needs or is inconsistent with their values.

I think that this approach should be adapted to the contemporary social condition. We do not need to choose only one of the scenarios, the one that wins the most widespread consensus according to the majority rule.[14] We should not approach the problem with only the "best scenario identification" bias in mind. Instead, we need to identify and design a number of scenarios, keeping in mind that we want to give each of them a chance and a place to be applied. Thus, we can offer different strategies for dealing with the mobility problem. We are not going to speak about the "mobility problem", but we will be concerned with the mobility *problems* of diverse people and groups. Instead of trying to synthesize the different problems in one huge and unsolvable one, an effective plan might be the one that seeks solutions for each of these issues and the diverse groups of people. In other words we must implement each scenario as much as it is possible and make different plans for people expressing different needs.

The designing of scenarios is a product of collective creativity which includes the creativity of the leaders and of the designers, but also reflects what citizens think … and they do not all think the same. Thus the question is: is the best scenario, the best solution, the best practice, etc., what we need to look for? Or should we rather strive to find a place for more coexisting scenarios, for applying more solutions and to continuously invent new practices?

"Cultural Creatives" vs. "Creative Class"

In this respect it seems opportune to argue against a current popular opinion, as, for example, maintained by the popular author Richard Florida (2002). Florida assumes that the progress of cities is based on the action of a so-called "creative class". As it is presented, this approach does not question the development model, the political structure and, mostly, it does not consider that real creativity emerges also from

---

[14] In the common sense, most people take for granted a basic idea of contemporary democracy, i.e. that the opinion of the voters' majority wins the right to be implemented. It is not necessary here to enter a philosophical discussion about this idea and e.g. quote the celebrated pages by Tocqueville (and Mills and others) who warned his readers not to confuse democracy with the "tyranny of majority". Tyranny of majority is generally rejected and considered against democratic practice, but the reaction to avoid this degeneration or misinterpretation of democracy has mainly been focused on reinforcing the possibility of overthrowing the government and on giving permanent voice to the opposition. The possibility to safely and routinely overthrow a government was a crucial problem at the times when the theory of liberal democracy was elaborated. Two different and yet plausible approaches are instead usually overlooked. First, we do not take into consideration the possibility of ranking and listing decisions according to the majority required to be made. For instance, simple majority is enough to deliberate when the effects of the decisions are easily reversible and there is a possibility of changing one's mind in a reasonably short time; a supermajority or qualified majority is required for decisions whose effects would last forever, like in the case of the construction of heavy infrastructures. Second, we do not consider the following possible strategy enough: instead of choosing and adopting the single solution, we should operate in order to create opportunities, in the political and geographical space, for the co-existence of more alternative solutions and lifestyles (Poli 1994c; also published in INSEE Paris Sorbonne 1994).

marginal social areas, and is often outside the economic establishment. The creativity of Florida's creative class is that of a chess player, restrained by limited rules and spaces. It is not the creativity of a pioneer exploring new frontiers, nor that of he who re-draws the chessboard substituting squares with triangles or, even more, messing up geometrical figures. So, instead of Florida's "not-so-creative class", we prefer to speak about "cultural creatives," that is, people and social groups who would be prepared to solve their problems – including mobility, as well as housing, safety, employment, etc. – in a radical alternative manner (Ray and Anderson 2000). If we really recognize and encourage this spontaneous creativity, then a new urban culture can arise from the ashes of decaying modern cities. In several European and American large cities' neighborhoods, we can find some of these social groups trying to adopt new life styles – including reduced mobility patterns. In Europe, the expectation of residents to implement new life styles locally is less likely. It happens mainly because the governments are more centralized and there is not a tradition of establishing new communities beginning from scratch. However, the geography and the society of middle-sized European cities would allow for the implementation of new environment-friendly projects concerning transportation. A high level in quality of life, in civic participation and in average income help this process.

The reference to Florida is also helpful because it is important to point out that he focuses mainly on larger cities. Considering the social and physical decay of major urban areas and the high environmental impact of transportation, we can find better solutions in examining the revitalization of middle-sized centers, which would include many European cities. These cities could well become innovative hubs. Their dimensions are also the proper size to implement a more direct democracy that is threatened by the loss of local power *vis-à-vis* the growing importance of national States and Federations whose territorial organization is typically hinged on their metropolises. If we assume that: (a) democracy is a value *per se*; and (b) that it is also effective in promoting development and wealth, then we cannot help but consider an optimal city/community dimension as one where democracy and participation can adequately apply. Technological and planning choices ought to be a consequence of these considerations.

Chapter 7
Northern Virginia Transportation Authority "Trans-Action 2030" Plan: A Case Study

Any fool can make things bigger, more complex, and more violent. It takes a touch of a genius – and a lot of courage – to move in the opposite direction. The significant problems we face cannot be solved at the same level of thinking we were at when we created them.

Albert Einstein

Abstract A few years ago the Northern Virginia Transportation Authority approved the so-called Trans-Action 2030, a transportation plan that included the Washington, DC metropolitan area. This plan is worth being analyzed because it reflects the contemporary state of the art, the current beliefs, and unfortunately, also the desperate invincible conservatism. The plan is clearly the output of the most advanced approach to transportation by civil engineering. Social scientists were employed only in marginalized roles and social research was almost completely neglected. As long as the planners have acted in the civil engineering discipline, they have been consistent with their methods and have proven high-level professional skills, no matter how we may have argued their opportunity. When the planners tried to apply social research and to face environmental concerns, they demonstrated a humiliating incompetence and the plan lost credit. The analytic description of the plan and the detailed criticism from social science and environmentalist approach points of view, is carried on in order to introduce the following chapters in which I propose how professionals other than civil engineers can deal with traffic problems and how different solutions can be employed to mobility problems in the future.

Keywords Transportation planning • Metropolitan areas • Social science • Environmental concerns • Decision-making criteria • Multimodal transportation systems • Communication tools

C. Poli, *Mobility and Environment: Humanists versus Engineers in Urban Policy and Professional Education*, DOI 10.1007/978-94-007-1220-1_7,
© Springer Science+Business Media B.V. 2011

A Commentary to an American Transportation Plan

In the previous chapter I introduced some planning issues and raised some questions about how to face the mobility problem. In this chapter I will analyze an example of the intrinsic conservatism in traffic policies, namely the Northern Virginia Transportation Authority's (NVTA) plan, called Trans-Action 2030.[1]

The NVTA Transportation Plan, released in Fall 2006, (Northern Virginia Transportation Authority, 2006) is worth being analyzed for many reasons. First, the area is a major national location because the Washington, DC area is a portal to the country for foreign visitors. The quality of life and the urban policy standards must be maintained at the highest level possible. Moreover, the population involved is significant and the growth rates previsions are exceptionally high. From another perspective, the plan is interesting because it is very well elaborated from a professional point of view. It could be used as a handbook in education and training of transportation planners. It reflects the contemporary state of the art, the current beliefs, and unfortunately, also the desperate invincible conservatism.

The plan is clearly the output of the most advanced approach to transportation by civil engineering. This is apparent both when you read the real goals pursued, beyond the rhetoric of the political discourse, and when you look at the research done to complete it. We might also ungenerously add that the planners clearly lacked any social science language as they make a number of very naive statements. For instance, in the "Resolution", it is emphatically stated that the "Authority shall, on the basis of a regional consensus, *whenever possible* (italics added), set regional transportation priorities for regional transportation projects". Now, from a political decision-making point of view, you cannot leave this statement so undetermined and vague. A properly designed planning process should include some clear statement on how to make decisions. You should know and be able to say how to limit possible local veto powers, and be able, as well, to define the limits of local consensus vs. general interest. Because this is clearly a difficult issue, if it is not possible to have anything precise to say, then it would have been better to say nothing.

If this is merely a naivety, a few lines below, the NVTA planners openly drop their masks. Although it seems as if at first they had been trying to conceal their straightforward technical training throughout the hundreds of pages of the report by conceding to several social instances, still, eventually they happily declare that the

---

[1]The fact that I have first introduced some of my ideas about how to plan and only subsequently I am passing to a critique of an existing plan, is a deliberate methodological choice, rather than a disorganized pile of disjointed ideas. I have elaborated my ideas in three decades of intense academic, professional and political activity. Hence, it would have been hypocritical to maintain that I had derived my ideas just from reading a few (or many) texts or from analyzing the last plan, the last report such as the Trans-Action 2030. My ideas have developed on the basis of a sound and serious methodological study and then have renovated thanks to experience and more readings all of which I have approached trying to avoid biases and keeping my mind open to new evolutions. Of course, I have not succeeded to get rid of my biases. Indeed, too much reading and theoretical study might have create more of them. So what?

"*regional transportation policies ... be guided by performance-based criteria such as the ability to improve travel times ... and to move the most people in the most cost-effective manner*." What does it mean to "improve travel times"? A social scientist, in general, should be a person with a certain sensitivity to how different people perceive the quality of a service, and would be prepared to argue at length about such a simple statement as "improve travel times." In the NVTA case, apparently it simply means to make travel quicker, as is proved by the last part of the sentence, which also states clearly that the goal is "to move the most people possible!" But, to make travel quicker, we can either increase speed or reduce length. Isn't this an option too? In the first case, all energy-saving and environmental considerations are secondary. Besides, does everyone agree that urban life and environmental quality are directly related to moving the most people the quickest way possible? Of course, someone would endorse this position and – why not? – with good arguments too. Others certainly would not. A major planning activity should have been to open a serious discussion on this critical point. If the authorities and/or the planners did not want to discuss it, at least the plan should have included the definitions of objectives, and should not have left this very crucial assumption totally unquestioned.

The Pariah Social Scientist

Notwithstanding their incapacity to detect crucial social and political issues of the planning process, a paragraph below in the report, the planners announce they want to know what people think. It seems they are not really convinced about allowing the opportunity for this to occur, but they must respond to the contemporary claims for more citizens' participation. So they make sure to propose a "*scientific telephone survey of 1,263 ... respondents, the participation of hundreds of citizens at seven community events, a public open house and hearing, and information dissemination through newsletters*."

There are three serious concerns about this poll. First, the study was commissioned to a consultant who was not part of the planning team. Thus, the communication process was not directly from the citizens to the planners. It was just an attachment to the planning process and was not an intrinsic part of it, as it should have been. The consultant did not have any voice in designing the communication process nor in stating his findings in the successive phases. The position of the sociologist in the planning structure and within the professional team was decided by the engineers and/or planners who controlled the plan's direction. We optimistically assume he was really a professional and an adequately educated social scientist, rather than a statistician or second rank engineer in a sociologist disguise. Because some engineers believe that social research is useless, they are likely to be driven by the supreme self-confidence that anyone can do social research and that it does not require any special skills. Second, I did not find out how much the social analyst was paid for this research, supposedly intended to bring information from the bottom-up. I suspect that the amount of money used to inundate the citizens with information from the top-down – see "*information dissemination through newsletters*" as mentioned in the second part of the sentence reported above – was more expensive and received much more attention than the – supposedly bottom-up – information the analyst gathered. Finally, I think that serious social research on the needs, preferences, and attitudes of the citizens, should have used more sophisticated survey techniques, similar to the ones that Companies customarily adopt in market analyses. For instance, an analysis that would have identified, quantified, and grouped citizens according to their preferences and attitudes – e.g. a cluster and/or factorial analysis, multidimensional scaling, etc. – would have provided much more workable information to the planners. Of course, only if they really wanted to use it, and if they knew about the availability of this kind of survey tools. It is not too demanding to call for high-level social research in such an expensive plan for such a crucial area of the country.

The planners, who hired a consultant to complete the scientific survey, wanted data "*on transportation priorities of Northern Virginia's citizens and their willingness to pay for additional transportation projects and services*." But they framed the question in a way that clearly suggested they already knew what the priorities were. The planners asked what the citizens were available to pay for additional projects.[2] What if they had asked how much money the citizens would be happy to save if no new project was approved, or offered citizens a choice among different tax and projects options? By the way, in the early 1990s, in Portland (OR), citizens were allowed to choose, in a referendum, between a new urban rail and a new road. It is absolutely clear that the NVTA plan's goal was merely transportation growth. It was as if someone were asked how much they would pay to have an orange, instead of being asking "what do you prefer to do with the dollars you own?" and being offered the choice between apples, pears, oranges, going to a movie, a hot dog, or a cheese cake.

The Real Decision-Making Criteria

A few pages below, the planners declare that "*extensive transit, bicycle, and pedestrian Level of Service (LOS) analyses*" will "*complement the highway LOS analysis*." They seem quite proud of having been the first to adopt these methodologies in Virginia, and among the first to apply them in the country. We all welcome the new methodology that by the way should more properly be called a technique. But it still seems incomplete because these analyses do not contribute directly to projects, and

[2]"Willingness to pay" is a well established methodology and easy-to-apply techniques are available. It is also highly questioned if adopted as a decision-making tool because – as they are Cost-Risk-Benefit Analysis, Social Impact Assessment and many other evaluation techniques applied to social issues – they may shed light on the problem, but do not guarantee any neutral solution and rather are intrinsically bias-loaded (Shrader-Frechette 1985). The passive application of this technique proves once more the intellectual laziness and the shallow methodological training of the Trans-Action 2030 plan's overseers.

instead, they just "complement" the pivot analyses still referred to as highway construction, improvement and maintenance. Nonetheless, every long journey begins with a single step and, at least, NVTA made it in the right direction.

Unfortunately, a few points later, we crash into another contradiction related to the "network-based performance evaluation criteria" that "*include:*

- *Provide an Integrated Multimodal Transportation System*
- *Improve Personal Mobility*
- *Improve Personal Accessibility*
- *Improve the Linkage between Transportation and Land Use*
- *Protect the Environment*"

This list is interesting as it reveals how the planners really rank their goals. Moreover, in the first point, the planners declare that they want to provide transportation facilities in general. They explain why in the following points. A consequent reasoning would have inverted the order of the points listed. Would it not be more sensible to state that, because they needed to improve personal mobility and accessibility, they would provide an Integrated Transportation System? This subtle inconsistency reveals the biases of the planners. Another contradiction pops up when they list both improvement of mobility and accessibility. If we think it is acceptable that both should be improved, then it would not even be necessary to say it. Planners should have said whether they intended to give priority to accessibility or to mobility, because this is both a real problem and a possible choice. All the discourse about settlement density pivots on this concept. The way it is stated in the Trans-Action 2030 plan clearly shows the planners' lack of self confidence. Nevertheless, just to make it clear, after all this tergiversating about accessibility, environment, multimodality, LOS and so on, they end up saying that NVTA agreed to "*a long-range transportation plan that will improve mobility*" … and that's it. Exit the clowns, enter the acrobats!

Moreover, the first two points are likely to require a big deal of construction, while fulfilling the following ones would entail mostly organization, some funding, and mainly some serious research. Last and, apparently least, "protect the environment" is mentioned, almost incidentally. The question is then: why and how to protect the environment? We all agree about it, of course, but it is listed as the last of the goals and performance evaluation marks. Should it not rather be a major condition or a burden to be fulfilled by the Environmental Impact Analyses? For at least 40 years, when speaking about traffic, environmental issues have been, at least officially, a major concern. Instead, according to what is reported in the Trans-Action 2030 plan, it reads that the planners assume that everything listed above must absolutely be done, just keep the environmental impact low … if possible. The option that road construction will be discarded due to its high environmental impact is clearly never considered. The performance evaluation criteria require the building of new physical structures and improving personal mobility. Because the building of infrastructures has apparently already been decided on, then it is absolutely superfluous to include the environmental protection criterion. It is mentioned merely as the last point, but could as easily have been left out. Could anybody seriously claim that we do not care at all about the environment impact of a construction? So, what for stating that?

This tricky inversion in the list is also demonstrated when the planners emphatically state the "Vision, Goals and Strategies". They claim that "*In the twenty-first century, Northern Virginia will develop and sustain a multimodal transportation system that supports our economy and quality of life.*" A logical argument would have declared clearly that Northern Virginia will pursue a better quality of life and, in that case, it will pursue economic growth if such is likely to contribute to its quality of life. Now, let us define what we do intend with quality of life and see how and if transportation might contribute!

How Would the Engineers Respond to These Criticisms?

Having some experience concerning how the engineers involved in transportation planning think, I can anticipate their replies to my comments. They would say that they were commissioned a Transportation Plan, so their duty was to provide and ameliorate the transportation supply, incorporating the input from the urban planners. As a matter of fact the Northern Virginia transportation planners received the following quite clear input: "*Within the next 25 years, Northern Virginia is expected to continue to attract highly educated professionals as the area absorbs approximately 651,400 new jobs, or more than half of the new jobs expected to come to the metropolitan Washington region. It is also projected to attract 918,500 new residents or 56% of those expected to relocate to the Metropolitan area. Today Northern Virginia is home to 2,164,700 residents and 1,238,900 jobs.*" Now we should raise two questions. First, is this growth good or bad? The answer should be left to the urban planners and/or to the political decision-makers rather than to the transportation engineers who typically claim that they are used to receiving the inputs and to providing solutions. However, sooner or later, the question ought to be answered and cannot be left suspended. Even worse, we cannot let engineers alone answer such a crucial issue.

Therefore, the second question goes: Should the transportation planners take a position about these previsions? I would definitely assent because honest analysts would seriously take into consideration the priorities expressed by the planners themselves. Certainly, they might possibly conclude that it was not possible to provide the same level of mobility to newcomers without harming the environment, wasting energy and without guaranteeing the same quality of life according to the traditional, given parameters. But they are required to prove it and to prove that they have tried to find an alternative solution. Another question is: Should engineers be asked for their opinion too? My answer is again a straightforward "yes" because they can provide valuable information about possible solutions and about the tools available to solve some distinct problems. In this case it does not matter how biased engineers' opinions may be, because the questions presume a broad array of diverse answers that are not hinged on the mere construction solution. Moreover, we do not know anything about the people who will move into the area. Depending on what we are planning to provide for them, we may have an influence on the characteristics and quantity of the future population. The plan does not say anything about these issues. The planners would keep claiming that it is not their business, and someone else should have taken care of the population growth and environmental impacts. The "banality of evil" is apparently applied to mobility planning and to the corporatist bureaucratic approach! Another common reaction to my criticism would be: "the situation is so compromised that something must be done immediately." This is certainly true. But an emergency should not push us to panic, unless that panic is deceptively manipulated by those conservative ones who want to build highways and nothing else. In fact, urgency should not imply a refusal to think about long-term alternatives, while at the same time trying to tackle present problems.[3]

Might We Have Reached the Limit?

The NVTA deplores the dramatic deterioration of the roadway system between 1999 and 2005. Thus, it immediately calls for more investments. Why not? But three other possible options are not taken into consideration at all. First, the dramatic deterioration may be the result of the excessive number of existing roads and of their overuse. Why don't we take into account seriously the hypothesis that we cannot afford, economically and from an organization point of view, to manage the intense use of so many roads? Second, we lack serious research proving that tax payers' dollars are really efficient if invested in more road building and/or maintenance. And why should Virginia planners adopt the previously mentioned network-based performance evaluation criteria if just a few lines below, they nonchalantly declare what they already know it should be done anyway? Third, are the citizens really willing to pay taxes in return of such a poor service?

After these methodological remarks, let us move to some observations about content. NTVA's planners write that they pursue the target of moving the most people, rather than vehicles, and that the projects should enhance connections among existing modes. This statement is made unambiguously. Nonetheless, research proves that the increased supply of movement opportunities will cause also car use to grow. Why? Usually, expert-planners (and most laymen-citizens) assume that building a new rail or bus service will shift commuters from cars to collective transportation means. A lemma of this (unproved) theorem is that this is good because collective transportation imposes a lesser environmental impact. The lemma is proven in some cases, but not always. E.g., mass transit works when there are no travelers for many hours of the day, and the energy costs are superior to the cost of few people driving

[3]The Italian Civil Code was approved in 1942 by the Fascist government while the regime and the country were falling apart. Nonetheless, it was a valuable enterprise and it is still successfully in force with the exception of a few obvious amendments. Probably, it was possible to elaborate such a momentous enterprise because of, rather than notwithstanding, the war and the political crisis.

their own car on the shortest route. Thus, we should calculate how much more environment-friendly collective transportation is compared to the use of cars. Even if we prove that mass transit is still more beneficial for the environment, the gain might be minimal. Likewise, the establishment of a mass-transit system requires a large and anything but flexible organization managed according to complicated bureaucratic procedures that are necessary when investing public money.

Finally, we must consider the possibility of induced effects on the overall mobility. For example, if there is a park & ride point near a new rail or bus station, I may decide to add an extra trip or to move an extra mile further from my job. The NVTA does not question the idea that more public transportation will just transfer people from their cars into buses or trains, and they endorse the idea that mass transit is *per se* an environment-friendly policy. In some cities, administrators and planners were conscious of this major problem. Unfortunately, this growing conviction did not yet entail a policy change.

Maps as Misleading Communication Tools

It is interesting to note how the NVTA's plan includes several maps which reported on road capacity, passenger loads, auto transit, highway performance and so on. Some of them are specifically questionable; for example, the one that identifies the traffic jams and the stop-and-go traffic spots. Indeed, they are elaborated with the evident goal of communicating that these bottlenecks should be eliminated by enlarging the roads. One could also claim that a reduction of traffic, namely a decrease in the number of people moving around, would reduce the bottlenecks and the stop and goes. These researches and these maps create the expectation that sooner or later something will be done to eliminate these specific stop and go situations. This message does not encourage people to find other solutions – such as taking a bus or finding a dwelling near the job, etc. – to avoid wasting time in bumper to bumper lines.

Nobody can stand being stuck in traffic jam. Car parks also concentrate exhaust fumes and increase local and global pollution. But also the construction of a beltway generates a high environmental impact on land and increases the average miles traveled and the average speed. Consequently we will have more exhaust fumes and more pollution. It is also naïve thinking that pollution would be dispersed on a broader area so that air quality improves: living in a high-density metropolitan region, we can just disperse traffic and polluted air … in someone else's neighborhood … while other drivers and more pollution will disperse in ours!

The publication of the maps, and the research completed to draw them, are powerful communication tools. The very fact that the planners have published those specific maps – hardly readable by the layperson, but bearing the halo of serious quantitative scientific inquiry – instead of the results of the sample inquiry on the citizens' preferences, confirms the mainly technical and mechanical approach of the planners. They are fascinated by those colored maps which do not say anything overtly about what they want to do, but only prove that they have collected lots of data, no matter how useful, or useless. Impliedly, those apparently innocuous maps stand for specific options. Drawing such maps is a lot of fun. It does not require any real thinking, just passively elaborating data. A psychiatrist would define the planners and decision-makers who adopt this coping strategy, as passive aggressive: they are passive because they apparently accept the requests of the people by mentioning environment and quality of life, bike and pedestrian facilities, and all the other wordy concepts, such as "multimodal" which they seem to like a lot. Then, they prove to be aggressive because they ignore all these concerns and eventually merely apply the usual methods, techniques and tools, as they intended to do all along. Actually, those apparently meaningless, colorful, and sophisticated maps, are powerful manipulative tools that create a biased perception of reality. They are even more powerful because the designers, lacking or refusing any social science methodological training, do not intend to consciously manipulate anybody. They strongly believe that the only possible option is the one inscribed in the maps. The most convincing preachers are the ones who really believe in what they sermonize.

Many Tools, No Policy

There are plenty of innovative solutions, fantastic new ideas, and congestion eliminating tools to be found in virtually any city in the world and on thousands of websites. The problem with traffic and, in general, environmental policies, is that the new ideas and tools enter into the planning process when it has already been designed according to the standard methods and has been conceived to pursue the usual goals. Administrators welcome these new bright ideas, provided they do not really alter the general structure. Minor traffic reduction projects match with the "ideology" of sustainable growth and distract people from thinking about a radically new development model. Governments, sustained by construction investors, cleverly appropriate some funding for several alternative insignificant projects which stand as nothing more than small de-tours on the road construction main course. Innovative designers and promoters are not going to make big money like the road and infrastructure constructors, but will still get some charity funds. They will also be gratified because they can apply what they have creatively elaborated. These brilliant and environment-concerned innovators have a technical, rather than a political culture. Thus, they hardly see the overall environmental problem, not to speak of the social justice, economic, legal and political problems. Being able to apply and implement their specific tool is all they ask for. Placating these honest and skilful, but politically naïve and intellectually lazy environmentalists, helps the government and construction lobbies to marginalize those who raise the political issues existing at the core of the traffic environmental crisis. Thus, political environmentalists lose the important support of their followers, who eventually prefer to believe in technology and keep politics in disdain. Therefore, the environmentalist potential for political change is defused. The transportation revolution does not simply mean to take a bus instead of a car. It means centering the political debate on the necessity to shift dramatically from a technical to a political culture, and from an economic development-oriented policy to an environment-concerned one. We should begin transportation planning with different ideas and tools – e.g. neighborhood schools, retail service distribution, pedestrian areas, etc. In general, soft and social actions, prior to physical planning, should be applied to organize a transportation system which might occasionally even include highways and mass transit structures.

Do Not Blame the NVTA Too Much!

Now, we cannot specifically blame the NVTA too much for all this. In fact, most of the transportation planners all over the world do not even think about considering the aspects I have highlighted here. Otherwise, there would be no need to write this essay. Nevertheless, because all over the world traffic congestion and city environmental inadequacy are rising to disastrous proportions, we are likely to have good reasons to believe that most transportation plans all around the world are simply wrong. If the next question proposed by the disappointed transportation decision-makers were, "what would *you* have done if you were in our shoes?" my honest answer would be, "something not too different." Indeed, I have had enough direct experience to know how difficult it can be to modify bureaucratic behavior. To a certain extent I even sympathize with the Northern Virginia decision-makers and planners.

There are two reasons why I would be unable to act much differently from the Northern Virginia planners. The first is that the eternally complaining citizens speak the same technical language of the planners, and expect a solution from the traditional engineers' methods. Then, of course they protest and complain because of the constant failures. Most citizens are not yet prepared to see or to try different solutions and innovative approaches. Consequently, if decisions originate by a direct democratic involvement, it is more than likely that a majority of the population would endorse conservative solutions over more innovative approaches. Respect for the people's will means that we must foresee and accept a conservative victory. On the other hand, the very same consideration about the will of the people should include two additional aspects: the first is that more information is necessary to give citizens the possibility to vote on real new options. But it is difficult to spread information on new projects and plans that are unusual, not yet applied, and not supported by powerful lobbies. Then, we need to protect the interests of minorities and allow them the opportunity to be represented, and for their opinions to be heard. It would be opportune and democratic to find these areas and the conditions to implement new projects and propose fresh ideas.

The second reason I would not have acted very differently from NVTA is that citizens are conservative when they think about transportation policies because they lack information about possible alternatives. It is easy for the layperson to evaluate a single project. It is much more difficult to penetrate the meaning of a complex planning process. That is why a composite plan, as opposed to a traditional one embellished with some minor environmentalist features, is likely to lose. One of the goals of the planning process should be to provide more information and to elaborate on diverse options. This does not generally happen because we lack a culture where transportation planning is approached from a social science perspective. Thus, environmentalists – who are the most critical of the current situation in transportation – do not propose anything really different from the transportation business as long as they face them on the same battle ground of the technical culture.

Conclusion

Traffic is not the pathology, it is the symptom. The real disease lies in city organization and in a development model which consumes natural and human resources! If traffic were pneumonia causing a cough, you would not hope to heal it by sucking more and more cough lozenges!

All that I have reported about Trans_Action 2030 Plan can apply to most European and American Transportation plans. During the time I spent at Johns Hopkins University Institute for Policy I had the opportunity to review many other traffic plans released by state and local governments in the US and in Europe. Notwithstanding my criticisms about mobility policies, we must admit that something is on the move. Many governments and many plans have adopted a language elaborated in some progressive academic milieu. There is a sincere concern about environmental problems so that politicians, scholars and professionals – including some civil engineers – try hard to move away from an unsustainable situation.

Chapter 8
Ethical Aspects in Traffic Planning

Never forget that only dead fish swim with the stream

M. Muggeridge

Abstract The urban traffic problem embraces ethical evaluations regarding environmental and health risks, environmental responsibility for global warming, social justice, and economic justice concerning the appropriation and redistribution of public monies. It also concerns an epistemic analysis regarding knowledge, science and the use of professional skills and competence. In Europe and in the US urban settlements and mobility patterns are significantly different and so should the consequent traffic policies be, especially those regarding short term mobility solutions. Particularly in the US, mobility policies may have diverse impacts on social classes and increase discrimination. A possible shifting from traffic policies to housing policies may have redistributive consequences and cooperate in equalizing income differences.

Keywords Ethics • Traffic planning • Urban settlement • Public expenditure • Social classes and mobility • US and Europe • Housing policies vs. mobility policies • Risk assessment

Ethical Aspects in Traffic Planning

The urban traffic problem embraces ethical evaluations regarding:

- environmental risks,
- health risks,
- environmental responsibility for global warming,
- social justice,
- economic justice concerning the appropriation and redistribution of public monies.

C. Poli, *Mobility and Environment: Humanists versus Engineers in Urban Policy and Professional Education*, DOI 10.1007/978-94-007-1220-1_8,
© Springer Science+Business Media B.V. 2011

It also concerns an epistemic analysis regarding knowledge, science and the use of professional skills and competence.

Mobility and the mobility-related technologies generate a number of health risks for the population. The most perceptible of them are, of course, air and noise pollution. The psychological stress caused by congestion and by ceaseless commuting is another threat that urban traffic poses on human health. Responsibility for global warming is also a good reason to reduce mobility and change the current development model.[1] A different traffic policy would significantly contribute to a decrease in emissions. From a utilitarian perspective – i.e. thinking in a cost/risk/benefit analysis scheme – the ethical problem is synthesized by the trade-off between, on one hand the advantages brought by this type of mobility, and on the other hand the costs it implies and how they are distributed among the citizens.

The (dis)advantages can be both social and economic. From a social point of view, we may hypothesize that a further growth in physical mobility would not entail a significant development of public and private happiness. Before telecommunication diffusion, one could assume that it was necessary to travel and move to acquire more information and open the mind to a multiform world. Urban enlightened people considered it progress toward civilization to allow rural area dwellers access to the cities to educate them in the urban way of life, in freedom, and to be open to cultural diversity. In the pre-global era, people's personal identities were by and large defined by regional life styles and local cultures that prevailed over international and national cultures which at the time were available only to the urban elite. "*Stadluft macht frei*" (city air liberates people) was Max Weber's celebrated quotation, who could not have imagined the obvious irony inherent in contemporary city air pollution. In fact, living in the city and traveling from one city to another was an opportunity to participate in an international culture and to compare different habits, beliefs and behaviors. Consequently, traveling and moving were means leading to knowledge. The diversity encountered by the traveler was more intense than any diversity found inside their own community, especially if that community was rural and isolated. Today, the situation is radically changed: easy access to telecommunication, as well as cheap and easy physical mobility, allow people to restrict their contacts to only others who are like themselves, no matter how far apart geographically they are. This implies a new form of isolation. The consequence of an excess of communication and movement can mean avoiding differences rather than fostering learning from others. It is possible to travel to wider and wider virtual and physical spaces in order to meet people who think and act alike, thus avoiding one's neighbor's diversity. On the contrary, if we were to live in a community from which, to a certain extent, no exit were possible, we would more likely be forced to deal with difference and diversity (Poli 2009).

From an economic point of view, the situation is, on one hand, easier to change and on the other, more complicated. The advantages of a further mobility growth are

[1] There is plenty of literature on the effects of traffic congestion and pollution on human health. On the contrary, there is still a lot to argue, from an ethical point of view, about the trade-off parameter between mobility and economic development, or about mobility and global responsibility.

likely to approach zero if assessed in terms of the value added of all production facilities minus the activity directly connected to mobility. A production system that survives only because of its ability to move huge masses of people and goods proves its absolute inefficiency. The contemporary availability of technologies would allow production of almost everything everywhere, including most agricultural products. We are currently destroying much more than we produce. If we cut thousands of trees to widen a road, then sell the logs, and produce a million cars that poison the air, in public accounting the result is an increase of GDP, i.e. of what is still considered a synonym of national wealth. This dramatically applies to developing countries too. Many people and many companies make their living by destructive activities, and thus they contribute to GDP growth. An immediate change of this system would have catastrophic consequences on economy and it is not advisable, as much as a gradual – though rapid – conversion is necessary.

Social Justice and Mobility Patterns

Mobility-related social justice problems are:

- how the same mobility patterns may have different effects on citizen groups?
- how public expenditure can favor different social groups?

At first glance, air pollution, compared to other forms of social discrimination, seems to be more democratic since it spreads all over an urban region. We breathe the same air if we travel in a Rolls or if we drive a wreck. Nonetheless, especially in North America, the poorest communities often dwell in the most polluted areas and are exposed to higher toxic emissions. Typically they live in neighborhoods located near congested throughways and are more likely to suffer asthma, other respiratory ailments, and cancers.

Moreover, transportation systems are designed to facilitate access from the suburbs to the center, rather than serve the city dwellers who are – with some exceptions – typically the poorest. Today, the greater part of low skill jobs are in the suburbs. It is true that in a growing number of American elite downtown neighborhoods, it has become easy and fashionable to live without owning a car and have access to the highly qualified CBD jobs. On the other hand, the lack of public transportation hampers the low income city dwellers from having access to jobs located in the suburbs if they do not own their own cars. In some distressed areas, like Detroit, even buying food has become a problem for the poor. Food stores are not accessible to people who do not own cars, since most small retailers have closed and the few who have survived have raised their prices to respond to the elite market created in the gentrified neighborhoods. Consequently, we ask the poor to own old cars, causing them to add to pollution, literally forcing them into social misbehavior![2]

[2]See Chap. 2, where I argue that in international politics the poor of the world claim the right to pollute.

Europe and US Mobility Patterns: A Comparison from an Ethical and Social Point of View

In the US, mobility and social justice are more interrelated than in Europe. In American laws and in the common way of thinking, the right to free movement is often emphasized and is considered an aspect of an individual's liberty. In Europe, mobility and migrations are perceived as a necessity and as a sacrifice that should be avoided if possible. Unfortunately, this attitude, which would be very helpful in the current situation, is rapidly changing. Europeans still consider it a social right of theirs to dwell in their home town and have a job in the area. Americans are more receptive to moving from one city to another and even from one state to another, in pursuit of better conditions. In Europe the right to move and migrate is not customarily assumed as an essential component of individual freedom. European workers would share the opinion of some American scholars who think that this claim for more freedom in movement hides the workers' exploitation by those who want a very flexible workforce (Clark 1983).[3]

Recently, however, in Europe because of the illogical growth of car use, roads and public transportation, the citizens' perception of the labor market's daily-commuting behavior has enlarged to broader and broader metropolitan compounds. The European penchant for owning a dwelling in a home community, where the family may have been living for generations, also has a negative effect on mobility and is a hindrance to mobility reduction policies. In fact, this attitude makes people prefer long commutes as an alternative to leaving their home community and moving to another nearby town. On one hand this is a positive factor *vis-à-vis* community life conservation. On the other, it has devastating effects as people continue to plead for more transportation infrastructures that governments are more than happy to subsidize. Because of the decision-making procedure we have described in previous chapters, governments are more keen to provide new commuting and traffic facilities rather than to re-organize the job market geography.

University student housing policy is a typical example of Europeans' tendency to focus on mobility – rather than, e.g., on residence – to deal with most problems. The Italian case is probably the most instructive. Italian students choose their University as close as they can to their parents' home. That way they can attend courses and stay with their parents in the town and community they grew up in. Often this choice compels them to travel for hours and to waste time that would be better devoted to studying, college activities, and amusements. Those responsible for university policies respond to the student uneasiness by pleading with local and central governments for better commuting facilities and transportation infrastructures. The obvious option to provide more student housing is disregarded. For example, while my students at the University of Bergamo (Italy) were doing research about ways to

[3]This issue was quite popular among liberal thinkers and politicians during the eighties when some American States (mainly in the so called "Sun Belt") attracted investments by providing major fiscal incentives. It created the syndrome of the run-away industries and forced many workers to move from the traditional industrial cities.

reduce pollution and traffic, the University President contacted municipal authorities to request more parking space for students and faculty, near (if not in) the car-free medieval city center.

The idea that the right to daily commuting prevails over all other rights is quite an entrenched one. This presumed right to daily mobility – or the lack of consideration given to the right of the people to have housing near their work or study place – has favored the establishment of many small university seats. The presence of universities in small cities and towns – traditionally an unusual pattern in urban Europe – has been endorsed by local ruling classes for reasons of prestige. Initially the idea was – as it happens in the US – to entice students and qualified professors who, by living in the city, would have advanced their own local urban prestige and cultural life. Instead, the more effective lobbying activity has been the one focused on requesting money to make commuting easier. The students have chosen the university closest to home in order to be able to go back home daily. Faculties did not even think about moving over there. Incidentally, the money students save in housing is typically used up in the purchase of a car. This is another entrenched materialistic bias: buying a "thing" is more socially and personally rewarding than "wasting" money for a service such as paying rent.

In most European countries housing problems have been transformed into mobility ones. Americans admire Europeans for their mass transit facilities. This admiration is justified only as long as it is presumed – although somewhat dogmatically – that public transportation produces less environmental impact than do private cars. But the relevant funds invested in transportation facilities make the change of mobility patterns more unlikely. Mass-transit unions and the management of transportation corporations – most of them transformed into publicly owned companies, that operate with market criteria, from public bodies as they used to be until a decade ago – do not have any interest in reducing commuting. Thus the coalition of these two strong vested interests to keep the *status quo* results in an invincible barrier to change.

In low-density US urban settlements, people are forced into frequent use of a car which entails high air pollution levels. The per capita emissions are superior to the ones registered in Europe although they are more dispersed.[4] The American settlement pattern has been harshly criticized – in my opinion more than it deserved – because it does not provide a community feeling similar to the one often experienced in Europe.[5] However, if we take into consideration the arguments presented about

[4]Generally, in environmental policies, it is taken for granted that dispersed settlements produce an overall higher environmental impact compared to densely populated areas. This idea has been challenged by some researchers (Neuman 2005; Holden and Norland 2005; Moos et al. 2006). This ongoing controversy is worth being taken seriously because in the next decades, the use of the Internet will eventually be introduced into the design of new urban settlements. Also, on the same theme, it will be important to consider the walled communities issue as a future form of urban settlement connected to the social, psychological, technical and architectural assimilation of web navigation.

[5]We must admit that the idea that European neighborhoods host real communities is today more myth than reality.

European commuting patterns and people's identification with the community to which they belong, we can claim that in some respect the American situation is more flexible. Thus, it may be easier to intervene to reduce overall mobility.

Nowadays, Europeans use cars and/or public transportation as much as Americans do, but they do it with different motivations. The average mileage driven by Europeans is lower than that of Americans because of the higher density of settlements, but the miles driven in daily commuting trips are nonetheless quite significant. Europeans drive less than Americans for daily activities such as shopping, visiting friends, leisure, etc., but not necessarily in travelling to work. European governments indirectly subsidize commuting via public transportation: the fees collected by public transportation companies still cover less than 40% of the costs, although EU regulations require this quota to grow. European mass transit companies are very powerful lobbies that manage substantial private and public resources, provide jobs and build infrastructures. These political and economic lobbies push the idea that public transportation is the only possible solution to traffic congestion and environmental problems. Should we believe in this solution without any research or critical argument?

Public Policy and Transportation

Mobility is also a problem of public policy as it implies:

- public money redistribution among social classes
- a different treatment of social classes

During my research on traffic policies in the US, I ran into a research proposal on Transportation Equity, written by Carmen Morosan for a Baltimore Foundation. Dr. Morosan is a member of the Johns Hopkins International Urban Fellows Association, of which I am the President, and is now working for the City of Baltimore. The proposal was not unusual for its content, which is perfectly intelligent and to the point, if not particularly innovative. But it grabbed my attention because it proves – together with other research on the same topic – that there is a real concern about the social issues of transportation policies. The problem is that this genuine and somehow elaborated concern proceeds on its own without ever coming into contact with civil engineering transportation plans. Only when the transportation plan is ready, may the two groups converge and meet. But at this point only details can be modified, while the general structure is already decided. Instead, a meeting of the two research groups should take place at the beginning of the planning process. I will come back to this.

As reported by Carmen Morosan,

> significantly, the modern civil rights movements began with fighting inequity in transportation systems. One of the first such instances occurred when Rosa Parks refused to give up her bus seat to a white man in defiance of local "Jim Crow" laws. This and similar actions sparked new leadership around civil rights in general and, as part of that movement, around

> transportation equity. Title VI of the Civil Rights Act paved the way toward social justice reform in all federally funded programs, including environmental and transportation ones.

Since 1970, US legislation explicitly mandated that transportation policymakers must take into consideration the benefits to and burdens on low-income and minority populations resulting from the planning process. The social and equity implications of the transportation policy process are a crucial issue and abundant research has been completed in this field. Most of the research, however, is focused on better ways to help disadvantaged groups to transport themselves to jobs using mass transit. Only occasionally have researchers taken into consideration how to bring housing and jobs closer or, more generally, how to resolve transportation problems by reducing people's need to move.

The Baltimore researchers analyzed data

> on race, income, employment, location of jobs and residence, and availability of public transportation for the Baltimore Metropolitan Area. The data were obtained from the 1990 US Census, the US Bureau of Economic Analysis, and from various Maryland State Departments, including Transportation, Planning, and Human Resources. The study used detailed information on recipients of Temporary Cash Assistance (TCA) in Baltimore City (where two-thirds of Maryland's TCA recipients resided), as well as applying Geographic Information Systems (GIS) on Census data for low-income neighborhoods, to estimate the proximity of residences of low-income people to mass transit stops. It also looked at the location of employers with potential to hire low-income people as entry-level low-skill workers, along with the frequency of public transportation to these locations.

The research was definitely interesting, but as the sentence reported above proves, it gives a picture of the situation without focusing on the preferences of the citizens and it does not propose solutions apart from those already known. The recommendations offered were the ones that anybody would expect, and thus absolutely acceptable to the mainstream. When the researchers eventually made recommendations, they suggested:

- *Provide affordable housing for people to be able to live near their jobs.*
- *Stop sprawl to the outer suburbs and promote Smart Growth policies.*
- *Improve the existing mass transit system.*
- *Expand the existing mass transit system.*[6]

The first two points look promising since they call for houses closer to jobs. But the last two points are focused solely on improving and expanding the current mass transit system which is nothing new and not necessarily compatible with the first two points.

The research also included a study on citizen participation, the results of which were also interesting. *Per se* this research, as many others more or less well elaborated, provides some useful information and valid recommendations. It reinforces a

[6] I thank the Baltimore Urban League for having provided me with this interesting information and Carmen Morosan for having discussed it with me. It is possible to read this report and others on related issues in http://onlinepubs.trb.org/onlinepubs/circulars/ec050.pdf.

way of thinking that needs to be stressed. The problem is that the research is – like a lot of similar studies – absolutely separated from the transportation planning activity. Therefore, it offers generic recommendations, but will not influence the transportation plan designers when they really make their structural decisions. It is apparent that jobs and houses are to be created, or offered at subsidized prices, near existing public transportation structures. Nothing is said about possible plans to design a transportation system that encourages a "smart growth", e.g. reducing the length of the bus/rail itinerary and offering more dense public transportation networks.

Chapter 9
Education and Training of Traffic Professionals

Education is what survives when what has been learned has been forgotten

Burrhus F. Skinner

We don't see things as they are, we see them as we are

Anais Nin

Abstract We need a broader approach to cope with the whole concept of mobility. Social scientists should be encouraged to apply their own disciplines in thinking about the economic, psychological, anthropological, social, political, etc. components of the traffic phenomenon. This is not to say that they should single out these aspects from the general problem. As we have mentioned before, engineers are likely to do just that with a narrowly technical approach. Nonetheless, we commonly ask them for both interpretation of the problem and a complete solution based on that point of view, an error that should not be extended to other disciplines. Scholars in social disciplines should approach the mobility problem comprehensively. At first, this would likely be limited to academic research and education. In the long run, it could provide a new perception of traffic problems and lead to various new applications of social science research. In this chapter I discuss a report produced by economists that demonstrates how civil engineering has "colonized" economics and represents an argument as to how economists should approach transportation policy more directly. In the following paragraphs some clues are given about alternative approaches in other disciplines, especially in psychology. In the future we should prepare new university curricula for traffic planners who approach mobility policies from a social science and humanities perspective.

Keywords Education • Curricula and syllabi • Engineering • Economics • Social science • Psychology • Humanities • Economic development • Lobbies

C. Poli, *Mobility and Environment: Humanists versus Engineers in Urban Policy and Professional Education*, DOI 10.1007/978-94-007-1220-1_9,
© Springer Science+Business Media B.V. 2011

Introduction

Paradoxically, we can assume that traffic and mobility problems do not really exist since they are just one option to categorize some urban inconveniences. Consequently, we can propose a completely different classification of current urban malfunctioning and not even include mobility problems in the list. Uttering this paradox as a "real" statement would be too much even in this radical essay. I use the paradox for what it is, i.e. a rhetoric tool that helps in explaining concepts and situations.

In this section I am not going to propose university curricula and syllabi to educate and train new professionals who ought to compete with, and eventually substitute for, civil engineers in dealing with mobility problems. This essay is based on the belief that we need a broader approach to cope with traffic. I claim that humanists and researchers in social matters should begin thinking about traffic and other environmental problems with the "*forma mentis*" and in the intellectual frame of their own disciplines. They should consider traffic as an economic, psychological, anthropological, social, political, etc., topic and not as someone else's business to which they might just cooperate when asked to. In order to approach and solve traffic and other environmental problems, humanists and researchers should start with an analysis in their own discipline. In other words, they should begin thinking about traffic as a problem that their discipline can effectively approach. Then they should be brave enough to enter into somebody else's disciplinary territory which, by the way, is exactly what engineers dare to do routinely. This is the opposite of saying that they should single out psychological, sociological and political problems from the general traffic problem, which is what already happens in many professional and scientific communities and planning teams. We have journals in sociology of transportation, in economy of transportation, in psychology applied to traffic, and so on. On very few occasions, though, do they report a comprehensive discourse on traffic in the same way civil engineers customarily do. This is actually a real (though concealed) paradox: we should expect engineers to be the group of professionals who are less keen to search for an overall vision because of the fundamental technical origin and essence of their discipline. But, as a matter of fact, we all ask them for both interpretation and complete solutions.

I claim that scholars in social disciplines should approach the mobility problem with the intention to solve it in its entirety, that is, in the same way that civil engineers would approach it. At first, this approach might not be a very practical one and will be circumscribed to academic research and education. In the long run, it will help to open new fields of research and provide a new vision to humanists interested in addressing technical traffic problems. For example, the idea that traffic and mobility are basically psychological issues might seem really revolutionary, if not categorically crazy. However, it would lead to reframing – or removing – much of the intervention apparatus that has become built-in to deal with traffic and mobility problems. Because we cannot avoid everyday obligations, if a psychologist should be put in charge of leading a traffic planning team, it is obvious that they would ask business-as-usual specialists for help. Nonetheless, this approach would be crucial from a research and educational point of view as it opens a path to an original way of thinking.

Unfortunately, at this time a large part of university research in social science and economics is funded by private and public companies. Researchers are requested to collect money for their institutions in order to keep their positions. A growing quota of their salaries are paid by the contracts they win. As a consequence researchers are not encouraged to "waste" time proving some innovative, alternative, weird idea which is not likely to be funded.

Traffic and Economics: The Colonization of One Discipline by Another

At present, economics is the most qualified among social sciences to challenge the engineering approach. In a culture dominated by a utilitarian approach, economics has developed the tools to quantify almost all phenomena by synthesizing them in monetary (quantitative) terms. Thus, economics can give a general view of what is going on. If we invite economists to consider traffic mainly as an economic problem, they will try to solve it within their disciplinary paradigms. Economists have already been involved in traffic policies from various points of view. They are often asked to elaborate analyses to evaluate the opportunity of some public investment.[1] On other occasions they are invited to analyze the impact on the economy of investments in infrastructures and technology. This is a typical example of colonization of one discipline by another which is politically more influential. I have examined some reports prepared by Cambridge Systematics, Inc. for Business Members of American Public Transit Association, Washington, DC. The authors wanted to prove "*The Economic Importance of Public Transit*". They (legitimately) promote the importance of Public Transit … for themselves, that is, for the investors. The problem is that economists hired to calculate the effects do not elaborate any alternative proposals to tackle the transportation problem.

After having listed a number of economic developments, the 1998 report[2] concludes that:

- Transit capital investment is a significant source of job creation. This analysis indicates that in the year following the investment 314 jobs are created for each $10 million invested in transit capital funding.

---

[1] Most of the time, they are hired to "prove" the opportunity of a decision since their most commonly adopted tool, i.e. the Cost/Risk/Benefit Analysis, does not guarantee either neutrality or objectivity as authors like David Pierce, Aaron Wildavsky and many more have proved in a very rich literature.

[2] The same approach adopted in the 1998 report is substantially confirmed in 2010 Cambridge Systematics' answer to the question: "Why integrate economic analysis in transportation planning?". See http://www.camsys.com/kb_experts_econ.htm.

- Transit operations spending provides a direct infusion to the local economy. Over 570 jobs are created for each $10 million invested in the short run.
- Businesses would realize a gain in sales 3 times the public sector investment in transit capital; a $10 million investment results in a $30 million gain in sales.
- Businesses benefit as well from transit operations spending, with a $32 million increase in business sales for each $10 million in transit operations spending.
- The additional economic benefits from the transportation impacts of transit investment in major metropolitan areas are substantial. For every $10 million invested, over $15 million is saved in transportation costs to both highway and transit users. These costs include operating costs, fuel costs, and congestion costs.
- Business output and personal income are positively impacted by transit investment, growing rapidly over time. These transportation user impacts create savings to business operations, and increase the overall efficiency of the economy, positively affecting business sales and household incomes. A sustained program of transit capital investment will generate an increase of $2 million in business output and $0.8 million in personal income for each $10 million in the short run (during year 1). In the long term (during year 20), these benefits increase to $31 million and $18 million for business output and personal income respectively.
- Transit capital and operating investment generates personal income and business profits that produce positive fiscal impacts. On average, a typical state/local government could realize a 4–16% gain in revenues due to the increases in income and employment generated by investments in transit.
- Additional economic benefits which would improve the assessment of transit's economic impact are difficult to quantify and require a different analytical methodology from that employed in this report. They include "quality of life" benefits, changes in land use, social welfare benefits and reductions in the cost of other public sector functions.
- The findings of this report compliment studies of local economic impacts, which carry a positive message that builds upon the body of evidence that shows transit is a sound public investment. Summarized in Sect. 6, local studies have shown benefit/cost ratios as high as 9 to 1.

We notice that there is nothing in this long list about the authentic and direct goals of a transportation policy. Clearly, the interest of the Association is to promote investments in public transportation and nothing else. They do their job and do it well. The problem is that nobody does other jobs that would also be necessary. The Association hired economists to prove the advantages of more public money invested in this economic sector. The economists, even more obviously, discovered that those investments in infrastructures help in promoting economic development and employment. Because all this is absolutely apparent, it is also unquestionable: no surprise everyone was happy … except that all this is supposed to be funded by tax payers who are not asked to choose among possible alternatives. Incidentally, we should add that only at point 5 is there some concern about savings in operating costs, fuel costs, and congestion costs. Nonetheless, if the investment in transit facilities will really boost economy as gloriously as the economists claim, it seems all too consequent that the overall traffic will increase. Thus making more investment necessary!

What is kept implicit in this report and in its predictable findings? Indeed, what is not declared is that:

- if investments in infrastructures favor economic development, they are good *per se*, independently from their original goal that was to ease mobility or to reduce traffic congestion or, even more in general, to improve the quality of life;
- investments in more infrastructures are good for the tax payers as they will receive a return in terms of more income which is an indirect, though unquestioned, indicator of wealth;
- (specifically in terms of transportation policy) more infrastructure is always good, and that there is no necessity to prove this;
- the only reason not to invest in it is that there is not enough money.

Therefore, if public money invested is likely to boost the economy, it would be wise for the government to move resources from other sectors, or adopt a deficit spending policy, in order to make more funds available for the construction of transportation facilities. With this very conservative approach, economists are "colonized" by the construction industry, refusing to take into consideration the many other factors which could be part of genuine economic research. For instance economists do not consider as assiduously as they should:

- an economic assessment of environmental damage; they do not add items to the lists of costs by including among the cost entries, items such as the loss of green land, pollution, energy consumption, etc.;
- research on possible alternative investments rather than the ones in heavy infrastructures.

Means and goals are dramatically confused. The viability of engineer-suggested transportation facilities goes unquestioned, hence economists are called to prove that those investments are also good for economic development. Economists should learn to question the dogma that more mobility necessarily improves people's quality of life and design their research to dissolve this possible doubt. A meaningful research cannot omit keeping the by-benefits of an investment (in this case production and employment growth) separated from the investment's real goal (improving quality of life via more mobility). Investment in transportation facilities should be a means to achieve the goal of a better quality of life. Instead, in the economic research quoted above, the goal is just economic development because the need to build infrastructures is never argued. Now and then, someone may question a single infrastructure, but no one dares oppose the idea that more infrastructures are better than fewer, whatever they are. In the assessment and decision-making procedures concerning the construction of mobility facilities, we should get rid of the persistent bias that confuses means and goals. The opportunity to invest in transportation infrastructures is reinforced by its positive effects on economic growth. If we assign too much importance to the positive consequences on economic growth – to the extent that it becomes the ultimate goal of the planning process – the project evaluation is clearly misled.

How Economists Should Approach Transportation Policy

What do we learn from this example? If economists want to win leadership in transportation policy, they should not crumble the problem into many pieces by responding to the requests of their employers. They should approach the traffic and mobility problem entirely by applying their professional culture and their disciplinary tools. It is possible that they might also find that, at a certain point, the construction of a bridge or of a light rail will be necessary. As a side benefit, the construction of transportation facilities may also favor economic development and employment, and for this reason it can be even worthier of being completed. Instead, the inescapable outcome of the current approach is that you cannot make a selection among convenient vs. useless transportation infrastructures since they are all convenient for a generic development policy.

The economist's approach should be research into the optimal allocation of resources, which might reasonably lead to a non-construction solution, or at least lead to minimizing construction and maximizing the utilization of the existing ones. Economists could study how to encourage – e.g. by tax policy, incentives, and creative public finance – a diffuse retail distribution in order to reduce traffic. They have the skills to evaluate and fix possible road tolls and other tools available in their discipline. Essentially they should pose the problems from an economic point of view from the beginning instead of accepting somebody else's approach. Economists might also conclude that infrastructure building is a policy apt to promote economic development and they can make this proposal if they take this as the objective of their action. If they want to take a lead in traffic policy, they should declare and evaluate the goals of the policies they propose, no matter what they are. Instead the current opinion regarding economic research in transportation is clearly described by Patricia Dowell of Cambridge Synthetics Inc. in the following quotation where she offers her opinion about the "Future of Economic Analysis" in transportation. Dowell writes that:

> There is growing pressure to be able to demonstrate the economic benefits and cost effectiveness of transportation investment. In many states, Governors and Legislatures are requiring that investments be tied to or measured against economic development goals. This trend of performance-based investment decisions has been further evidenced in the recent Federal Transportation Investment Generating Economic Recovery (TIGER) Grant under the American Recovery and Reinvestment Act (ARRA) legislation, which required a rigorous analysis of benefits as part of the application process. Indications are that the use of economic analysis in Federal funding decisions will expand to include the Transportation Infrastructure Finance and Innovation Act (TIFIA) program, the joint HUD-DOT-EPA Sustainable Communities grants, and other discretionary programs that may be funded in the next Federal authorization bill. Moreover, as funding for transportation continues to lag behind the investment needs, agencies are looking for new funding alternatives and revenue streams, including public private partnerships. Economic analysis plays an important role in identifying projects with the highest probability of attracting private investment and generating new revenue streams. In addition to project selection, prioritization, and funding, the use of economic analysis is playing a larger role in asset management, performance measurement, climate change, and pricing strategies.
>
> Cambridge Systematics has been a leader in transportation economics and we have a team of dedicated economists that continue to assist Federal agencies, states and local

governments, and business and economic development organizations in developing guidelines, tools, and applications for integrating economic analysis in transportation decision-making and policy.

As a matter of fact there is no other investment that is more capable of promoting economic growth than transportation funding. It involves construction, steel and other materials, vehicle manufacturers, land development and more. Transportation is traditionally the keenest to receive public funds. Because the economy and society rotate around transportation, if we want to reform economy and society we need to change the way we think and act about mobility and the transportation industry. Therefore, if economists and other social scientists were able and willing to participate in transportation research, they could contribute to a political revolution even more than to a scientific paradigm shift.

A Traffic Plan Designed by Economists

One may raise the question: "Why should economists design a plan if they have not been asked to?" The point is that economists do not think that they can design an unabridged mobility plan. Because they do not see themselves as leaders in the mobility planning process, no one asks them, nor do they develop any real competence in the subject. If economists realized that traffic could be their concern as much as it is the engineers', they would promote academic research in order to propose economy-based mobility plans and develop a new disciplinary discourse.

The economist's approach to mobility should be based on the following considerations:

- The construction of traffic facilities requires a considerable amount of government funding and, consequently, taxes.
- Taxes in effect, redistribute income among citizens depending on specific investments to which they are directed.
- The quota of "government-led vs. market-led" income redistribution is a crucial political issue.
- Economists should raise the following question: can government funding be bestowed in a different way in order to favor better mobility and/or transportation?
- Because building transportation infrastructures means that a significant amount of taxes needs to be raised, we can either reallocate taxes differently, or reduce them.
- In the case of them being used differently, we can study incentives that reduce the necessity for people to travel.
- Economic projects/models can be formulated by excluding or minimizing the hypothesis that new infrastructures are necessary.
- A typical question that economists should try to answer is, how much tax reductions for small businesses in retail distribution can boost this sector and reduce the necessity to drive to shopping centers or to invest in infrastructures to reach them by car?

- Governments can afford to reduce taxes because they will reduce expenditure for infrastructures.
- As a result there will be a shift in the people who benefit directly from public expenditure (namely from constructors to small business retailers), which is, by the way, a real libertarian/free-market approach.

The same research can be made into housing, schools, hospitals, and more. Eventually, the main assumption of the economists should be: if traffic is a form of inefficiency in urban life, how can we repair it by using our disciplinary tools and without relying on more construction solutions?

Other Disciplines

Specialists in single subjects should dare to deal with general and complex problems. Their attempt, based on different cultures, will lead to new original knowledge and to a new vision. Humanists and scholars in social issues could devote more of their time and energy to approaching the general problem of traffic, instead of focusing on the side issues that engineers assign to them. Humanists dealing with traffic should propose a new comprehensive vision of mobility policy that should become a competitor of the civil engineer's purely technical approach. The fact that the overall problem is approached from a different point of view will make a notable difference. "Applied" humanists and social scientists should accept this challenge and wage war on other professionals who are doing too much within their own competence. If sociologists approach the mobility problem holistically – and if they do it creatively – they may focus on issues such as community life, how people interact, the effects of moving around daily, and so on.

In the previous sections I have suggested a traffic plan led by a planner whose background was mainly in political science or in related subjects. We assume that this is still the most practical approach because planners are available and respected in the professional market, especially in the US the problem, as I have argued, is that the technical political discourse is such that even planners cannot really affect the civil-transportation engineer-led process. The main reason is that planners cannot exercise their influence like social scientists because humanists are not interested in dealing with problems that are unfamiliar to them. Training humanists to deal with traffic planning problems requires time because for a long time they have been marginalized in the political arena and they are not yet prepared for problem-solving in this and other subjects. Thus, rather than propose an immediate list of examinations that students in human and social science disciplines should take, we need to encourage scholars to begin studying traffic problems from scratch relying on their disciplinary backgrounds. As in Anais Nin's quotation introducing this chapter, they will fashion new, personalized views of the problems. They need to bravely explore lands they are not familiar with, and show the chutzpah (impudence) to challenge professional engineers who firmly believe they are the sole custodians of a revealed truth, at least in transportation planning.

The occupation of the problems by discipline specialists organized in politically and culturally powerful guilds, and the substitution of technicians with humanists should not be considered an absolute and uncontroversial progress. It was a consequence of a way of thinking that was first systematized by René Descartes[3] and became a characteristic of the modern era. Before modernity, humanists were in charge of framing problems and locating them in a wide perspective. If we go back to antiquity, we have several examples of highly civilized societies where political action prevailed over technical knowledge; in other situations techno-structure posed the problems and kept the society enchained to an immutable method to approach them. Most of the time, civilizations in the course of their history moved from the political to the technocratic approach.[4] Scholars from diverse disciplines can approach the traffic problem and propose themselves as leaders in problem solving. To do it they should reframe the problem looking at the aspects with which they are more familiar. This move is per se political and revolutionary.

Psychology and Mobility

The "Association of Applied Psychology, Division 13: Traffic and Transportation Psychology" has organized an International Conference on Traffic & Transportation Psychology. In the conferences, traffic psychologists virtually examine all aspects of human behavior in traffic … but neglect the general problem. If we take a look at the program of the 2008 Conference, we note that even the keynote speakers proposed a segmented vision of the traffic problem. Ray Fuller spoke about "Who drives the driver? Surface tensions and hidden consensus"; Ian Glendon was even more technical with a keynote on "Changes in brain structure of younger and old drivers"; Karl Peltzer examined "Road use behavior in Africa"; David Sleet dealt with "Integrating public health with traffic psychology"; and Floura Winston finally introduced "Opportunities and new challenges for impacting transport safety among children, teens and young adults".

The only keynote partly in the spirit of this essay was presented by Steve Stradling who approached the broader theme of "The challenge to transport psychology: How to get the world to burn less fossil fuel". This would have been a good starting point to challenge the usual approach and propose psychology – better say psychological culture – as a leader discipline and a practical culture with which to revolutionize the traffic business. If psychologists are put in charge of the urban transportation plan – or if they pretend to be in charge and develop research as if they were – they

[3] I apologize for this simplistic pop quotation about such a huge issue as the fundamentals of the scientific method. However, readers who know what I'm talking about do not need further clues. The ones who are not involved in epistemology are supposedly not interested in this argument or will be stimulated to read more about it.

[4] See Frijtof Capra's interesting considerations about the scientific method and Leonardo Da Vinci's science (2007).

will "see things as psychologists do". Their priority might be issues such as the psychological well being of drivers and city dwellers. Traffic psychologists conduct research on topics such as:

1. driver training and licensing,
2. driver impairment,
3. effects of fatigue,
4. characteristics of the aging driver,
5. perceptual and cognitive disabilities,
6. road user attitudes and behavior,
7. road user needs across the lifespan,
8. enforcement and behavior change,
9. driver support systems,
10. the psychology of mobility and transport mode choice.

A major investment in these researches and in the implementation of the plans elaborated by psychologist-led planning teams on these themes would bring a radically different perception, and consequently, a new approach to mobility policies. Because it has not yet happened, we do not know if, for example, analyzing road user attitudes and enforcing their behavioral change (point 8); or enhancing driver support systems (point 9) would have better effects on mobility than the customary construction of roads and light rails in relation to any dollar invested. If psychologists do not propose a general transportation plan, they will never learn how to play a key role in the subject and will remain happy to be sub-contracted by engineers and constructors who will leave them only the crumbs of their abundant meal. Only if they approach the traffic problem comprehensively will they have the possibility to introduce some real cultural (and political) change.

Conclusion

The same arguments I have brought forward for psychologists and economists can be used for sociologists and other scholars and professionals in human disciplines. Humanists can take the traffic problem and try to understand and solve it without any inferiority complex *vis-à-vis* civil engineers. In the beginning this would be just a new educational approach for student training. In a decade of thinking, researching, and teaching, we might be able to create a new culture and would be able to train professionals prepared to tackle mobility, transportation problems and other urban problems from a new perspective which would likely be more effective and environment-compatible than any current pseudo-solution.

Chapter 10
Planning Approaches

The world is not knowable and there is no truth ....
Just as you can't learn the taste of salt with your nose,
the smell of balsam with your ear, or the sound of a violin
with your tongue,
it's impossible for you to grasp the world with your reason.
"With what can you grasp it?"
"With your passions ..."

Isaac Bashevis Singer, "Zeidlus, the Pope"

Abstract Urban traffic plans are authentic political acts as they reallocate resources among taxpaying citizens and because they entail substantial change to people's everyday life. Nonetheless, traffic plans are not submitted to rigorous political control. Consequently, we leave it to technocrats to undertake a political function they are not supposed to perform. This brings injustice and inefficiency. Two questions ought to be raised: (a) should the plan process allow for the consensus of the minority and to what extent? (b) who are the professionals who can best advise planners in structuring the process? The question of consensus is a crucial one in traffic planning, especially when approached from an environmental point of view. Our value system, as it impinges on the environment, is a relatively recent construction. That is why it is difficult in practice to choose among conflicting goals. Two dangerous personalities, i.e. the enlightened despot and the passive-aggressive leader, often cooperate in urban politics and together they create disasters. The traffic plan should be a communication tool and we need professionals to use it. As a consequence, the social research involved will be sophisticated and will lead to a better and more democratic information-communication-education campaign. In this chapter I provide some ideas about how to approach mobility and will challenge the "more mobility, more wealth paradox".

Keywords Information • Communication • Education • Environmental justice • Decision-making • Urban policy • Planners • Politicians • Psychoanalysis • Social surveys • Citizen preferences • Scenario building • Plans vs. projects

C. Poli, *Mobility and Environment: Humanists versus Engineers in Urban Policy and Professional Education*, DOI 10.1007/978-94-007-1220-1_10,
© Springer Science+Business Media B.V. 2011

Introduction: Traffic Plans as Political Acts

Urban traffic plans are authentic political acts as they reallocate resources among taxpaying citizens and because they entail substantial change in people's everyday life. As we have defined contemporary cities as places in which you move around, mobility and traffic related policies play a crucial role in urban politics and city governance. Nonetheless, traffic plans are customarily elaborated, presented and intended (also by the citizenry) as if they were purely technical documents. Thus, in most of their elaboration phases, it seems legitimate that they are not submitted to a rigorous political control. Consequently, we are in a situation in which technocrats undertake a political function they are not supposed to perform. This repeated political abuse brings injustice and inefficiency. To restore a legitimate situation, we need a (traffic) revolution, political and cultural at the same time.

The Planning Process Approach and the Goals of the Plan

When transport engineering companies apply to a city administration to design an Urban Mobility Plan (UMP), they usually avoid avowing the plan's goals. If they do, they list very simplistic and predictable objectives. Most of the more general goals are kept unspoken although a discussion on the plan goals would be the crucial and most intellectually stimulating phase in a plan design and the only positive starting point. Someone would claim that comprehensive discussion about society's ultimate goals might be superfluous and time consuming. I believe that it is not. I need to add, though, that such a debate may be seen as politically dangerous to change-averse people. The time allegedly "wasted" in building a vision for the future is indeed rewarded with new ideas and solutions, and even with an easier and more legitimate decision-making.

When we begin dealing with urban traffic and mobility problems, as a first step we should devise a planning process that generates just and efficient decisions that have the potential to solve a good portion of the social problems inherent in traffic. In our current society, in which people feel that they have the right to be involved, decisions cannot be effective if they are not perceived as shared and just. The definition of a plan's ultimate goals should be the focus of political decision-makers (PDM) and of planners at the beginning of the process. PDM's and planners should invest part of their time studying all possible options, which can be numerous and conflicting.

By definition, the planning process is a typical political act, even though professional planners can contribute to improvement of the design of the process. At this point, two major questions ought to be raised that those responsible for the UMP must answer after thorough research and shrewd meditation:

(a) Should the plan process receive the consensus of the other political parties (the minority) and to what extent?
(b) Who are the professionals who can best advise them in structuring the planning process?

Some Clues About Planning Problems

Regarding point (a), I am perfectly conscious that the answer I will offer in the following paragraphs can be reasonably disputed. Even so, I want to stress that an open and preliminary discussion is necessary to make the plan's political and social actors aware of what they are really doing and what direction they are heading in. This discussion may take place openly or covertly, depending on the political situation. However, the political relations issue should be taken into consideration in order to understand the power and the institutional dynamics. We must consider the following basic points:

- The design of the planning process is a distinctive political act of a majority that is entitled to make decisions.
- The specific decisions taken in the following phases of the plan should require the citizens' and their representatives' involvement.
- It could be possible to claim legitimately that consensus is not necessary on behalf of the minority because the voters' mandate should be enough to entitle the majority to make a unilateral decision.
- I call for a broad consensus on each phase of the planning process and on each specific decision.
- I choose this option mainly for reasons related to political competition fairness; but also – somewhat cynically – for practical ones, since it is easier to face possible dissent politically and in an institutional frame, rather than coping with strenuous people protests and never-ending legal actions.

In designing the planning process the authorities apply the power legitimately committed to them by voters. But it is not reasonable for the majority to design the plan in its details and impose it on the minority, even though the majority is formally legitimated to design the process. A possible option is negotiating with the minority to agree about the conservation of the planning process beyond the mandate. In this respect the political content of the planning process design is even more emphasized. Such a long term agreement (i.e. lasting beyond the mandate) would become easier to achieve if the planning process includes the tenet that an agreement should be found on any incremental decision and to approve any major project. Because a UMP includes public works which will last for a long time – evidently much longer than the voters' mandate – a consensus broader than the simple majority seems necessary for a political ethical reason: you cannot transform the city and its territory forever with merely a bare Council majority vote, which might not even represent the majority of the citizenry. Before the environmental crisis, this was not a concern because there was implied widespread consent about building, developing land, and transforming natural resources. These days some high environmental impact constructions and decisions – such as the ones regarding mobility and transportation – are likely to require qualified majorities in deliberative bodies (for sure) and probably also referendums. The contemporary dominant value system and our democratic institutions are inadequate and inappropriate for guiding our thinking about most environmental problems. "This value system, as it impinges on the environment, can be thought of as a relatively recent construction, coincident with the rise of capitalism and modern science, and expressed in the writings of such philosophers as Francis Bacon, John Locke, and Bernard Mandeville. It evolved in low-population-density and low-technology societies, with seemingly unlimited access to land and other resources. This value system is reflected in attitudes towards population, consumption, technology, and social justice, as well as towards the environment" (Jamieson 2002: 291; Hirschman 1977).

Every decision having long term consequences (often perennial) on the environment is abusive and unethical if it is made only by a slight majority of the representative body or even by a slight majority of the people in a referendum. We can say that it is unethical because of the impossibility of reversing the decision once the constructions are completed. The libertarian economist Von Hayek claimed that before allowing a house to be built, the government should evaluate the decision much more carefully than when it funds soft activities, because the house will last forever while financial flows can be revised at any time. Utilitarian philosophers have thoughtfully contended as if ethics were, indeed, the most advanced form of opportunism. We will not enter this dispute. However, besides ethical considerations, we claim that it can be also convenient and efficient to design a planning process that requires an extensive consensus at each phase, in order to reduce the decision-making times. In the case of a UMP that does not consider any new construction – any perennial physical impact – I assume that decisions can be made incrementally by simple majority. Thus, we need to distinguish accurately between soft traffic plans and physical infrastructure building.

Conflicting Goals

Once the planning process and the plan's overall approach have been decided upon, we need to set more specific goals, or second-level goals. In the current social and economic situation of Europe and the United States, these second-level goals are likely to be related to a better quality of life rather than to economic development and infrastructure building. Thus, identification of goals implies more detailed considerations regarding the different opinions citizens may have about quality of life. A quality of life related to physical mobility can be pursued by:

- reducing traffic congestion,
- improving air quality,
- decreasing health risks,
- increasing personal income,
- increasing employment rate,
- enhancing traffic-related safety and security,
- improving accessibility.

Compatibility problems, different priorities and contrasting opinions are more likely to show among the citizenry when we declare more specific plan goals. All the diverse opinions must be respected even though planners will naturally have their personal ones. However, the planners' task is not to decide how best to proceed, having already chosen the solution they consider optimal or the best possible. This would be the technical solution that we have been criticizing from the very first pages of this essay. Planners should not make personal decisions by proposing solutions, at least at this phase of the plan designing. PDM's and planners are to design a plan that will be accepted by most people, as it is appropriate in a democratic country. Their goal is to build a sound and legitimate process which facilitates and allows easy and shared decisions. Planners must learn to consider a plan successful when they are able to obtain an easy and shared decision, although they might have thought of a completely different solution. In other words, PDMs and planners should not have or show opinions on the content and the solutions, at least at this stage of the planning process.[1] Planners apply their intellectual influence by introducing into the planning process innovative options and possible newly formulated solutions, or those unavailable before the beginning of the planning activity. Good planners are those who propose many new ideas, and then are willing to let the public evaluate and choose among them, possibly via the effective decision-making and just processes they have helped devise. The border between the political planning phase and the technical one is inevitably blurred, but the very attempt to define what is political and what is technical helps to improve comprehension of the actions.

Two Dangerous Personalities

Some social scientists have a hidden, unconscious personality. They should detect it and let it emerge at a conscious level after a series of psychoanalysis sessions. These social scientists – including planners – should be allowed to work only after having visited a shrink for some reasonable time. The hidden personality to be revealed is the one of the "enlightened despot," that is, the one who knows everything and wants to impose their opinion for anything less than the progress of humankind. Because this rarely, if not ever, happens, the wannabe-enlightened-despots live in a bivalent condition of frustration and self-congratulation. They are frustrated because they cannot apply their ideas; at the same time they self-congratulate, claiming that their ideas have not been applied because they lacked political support. This attitude is the opposite of an equally dangerous personality: the "passive-aggressive-political-leader" who is completely uninterested in knowing anything, the one who has neither opinions nor preferences, whose only goal is to lead the people with the same attitude of a shepherd who follows the sheep. These two narcissistic and self-centered personalities are fatally attracted to each other. In fact, they both find it hard to relate

[1] Gunnar Myrdal's studies on objectivity in social science were part of my basic readings during my doctoral studies, so that the problem of objectivity and neutrality has always been one of my major concerns as a social researcher.

with others so they build a context in which they do not need to truly interact although they work side by side. Their meetings are as frequent as the disasters they perform. In fact, passive-aggressive leaders renounce fighting for their ideas, their main interest being to keep the lead by doing what the most powerful – e.g., the powerful constructors lobbies – ask them to do. The wannabe-enlightened-despots will focus on their dream of the perfect plan which, of course, will never be implemented. Hence, they will blame the poverty of politics and politicians, the incapability of the people to understand their brilliant ideas, and eventually the fact that they have not been given enough power. The passive-aggressive leaders might see their power endangered if they hire planners who care about an effective political process and, moreover, aim at introducing new serious possibility for real change. But the most important psychological reason why passive-aggressive politicians refuse any interaction with the wannabe-enlightened-despots is that they are uninterested and incapable of doing so. On their behalf, the wannabe-enlightened-despots have the same communication problems. Eventually, the meeting of these two characters is the main obstacle to change and generates a very conservative political environment.[2]

The Mobility Plan as a Communication Tool: Polls and Professional Competence

I cannot guarantee that I will be able to design a planning process completely consistent with the ethical-political requirements I have outlined above.[3] Probably not, but the very attempt to adopt these guiding principles would help to introduce a different perception of the mass mobility problem in cities and would make change more likely. If we continue to apply the usual methods, and do not try to introduce a different way of thinking which will respect and include a variety of the attitudes and beliefs of citizens, we will be doomed to be stuck desperately in a traffic jam. Collective learning is a crucial aspect of the planning process. Thanks to the planning activity, citizens will learn new interpretative hypotheses and possible solutions that they could not imagine before. Planners and PDMs must play a crucial role in introducing new ideas, points of view, solutions to problems, interpretation. The decision-making process, however, follows a different course that must be thought about and designed separately. This is the main task of the PDM whom planners consult.

[2] I hope that the reader will not take too seriously my amateur psychoanalytical effort. I owe Crozier and Friedberg (1980) the image of the social scientist as an enlightened despot on which I have elaborated. Also Gunnar Myrdal was aware of this problem and treated it extensively (and seriously) in the methodological sections of his "*Asian Drama*" and in "*Value in Social Theory*". Ironically, he maintained that we should carry on an anthropological study on the social scientist community since we are not less interesting than new immigrants, prostitutes, problematic youth and so on as reported in chapter one.

[3] Even less can I guarantee that I need not see a shrink for some detrimental hidden personality of my own.

To open an effective debate on new ideas and their interpretation, it is necessary that the plans (or the planning process) are or become part of the people's everyday life so as to put the citizens themselves in the condition to understand and re-elaborate what they suggest. Legitimacy and consensus are crucial for the success of an innovative plan: a primary planning process goal should be the free expression of the citizens' preferences. Most of the time, in order to let the people freely and effectively express their preferences, planners and PDMs survey citizens' opinions by questionnaires or by public opinion polls. There are many methods of finding out what people think and none of them are either value-free or neutral. Thus, it is necessary to study carefully the technical approach and the political significance of polls. The professional in charge of the survey cannot be a statistician, expert in calculations. Computational expertise is important and helpful, but it is necessary to have sociologists, psychologists, political scientists, and others, to interpret the information collected and its explanatory limits. A multidisciplinary team is appropriate provided it is led by a professional (or even by the PDM) who is able to use someone else's competence. Today, the specific competence of someone who effectively knows what the others can do is one of the least available in the professional market due to the century old preference given to specialization. Universities should provide courses to teach what disciplines really do, can do, and should do. No matter how shallow and taxonomic these courses may be, they would provide students and professionals with a crucial knowledge about what others do, can do and should do.

Besides applying a sufficiently neutral and objective opinion poll, we should also expound the method, the content and the goals of the poll. This operation is useful for at least three reasons:

- it helps the planners and PDM to clarify – to the audience and to themselves – their objectives, what they are really doing, and where they are headed;
- it helps to advance the method adopted;
- it makes explicit several assumptions and therefore helps the consensus formation.

It is an educational procedure because it promotes discussion and increases collective and individual knowledge.

"For *your* information, I would like to ask a question". This witty quote by Sam Goldwyn explains how the surveys produced to find out the citizens' opinions and preferences are a two-way communication: we do not only collect opinions, but while raising questions, we also send messages, provide information about new possibilities, and emphasize some topics rather than others. Without hypocrisy, I must admit that the research direction, no matter how neutral and objective we try to be, is to a large extent determined by the plan's political leadership. But "to a large extent" does not mean "completely." It is still difficult to eradicate the bias that a proposition can only be false or true. To think that we can dramatically separate truth from fallacy is useless in contemporary social science. In human life, overlapping between true and false is normal and it represents a spur to change and progress. Thus, the plan should be something between a forum – a place to debate ideas and advocate for one's own interest – and a decision-making tool. Admittedly, I am simplistically exposing issues that have been deeply debated for decades in social science methodology. However, now and then, it is sensible to recall the methodological debate which has not yet been solved and is still open. The fact that we have not found a solution and that it is likely that there is no solution does not mean that we should give up coping with problems!

How to Survey Opinions: An Informational and Educational Communication Project

A few years ago I was contracted by the municipal government of an Italian city to research the attitudes of its citizens regarding the environment. The research goal, as I myself proposed, was to design an Information, Communication and Education plan (ICE) that would work as background knowledge for other, possibly future, plans, namely those for traffic reduction and cleaner air. The research included a questionnaire submitted to a stratified random sample of 1,000 citizens officially living in the city, from a population of approximately 200,000. The added value of this research was that 9 years before I had examined a comparable sample. Therefore, we could examine present attitudes and see how they had changed in 9 years. These two researches – and many others carried on in more than a quarter of a century of professional and academic research – are the cornerstones of many ideas contained in this essay. The last survey was conceived to be the ground work for a traffic plan in the spirit of this essay.

The city government – as most other city governments – made two fundamental mistakes and eventually failed to adopt a promising approach. Regrettably, the administration had invested a conspicuous amount of money in this research, but was not able to use it properly.[4] The first mistake was that the decision-makers expected that I would provide them with a series of actions and messages generically addressed to the citizenry so that the citizens would more easily accept a plan already prepared by technicians. They promoted only one real political action; namely, they negotiated decisions with retailers and craftsmen's associations, and some other vested interests. The planners – all engineers or bureaucrats – thought they already knew all that was necessary to implement the perfect traffic plan that they had already completed.[5] I suggested using my serious survey of the people's opinions and preferences that I had just delivered to the municipality. We could have designed

[4] Frankly, I do not regret that, once in a while, I was well paid for my research. I sincerely lament that the public money was not used effectively. This situation explains how change is difficult. You need to be able to commission the right research; then, you should carry it on properly and honestly; eventually you need people around you who can apply it effectively. If just one of these steps is missing, the entire process fails.

[5] This is a typical example of the encounter between the passive-aggressive politician and the wannabe-enlightened-despot, except that I honestly had no intention to act as the latter, but eventually I ended up being in that exact position.

the decision-making process and the specific decisions by taking into consideration all that we had learnt from the expensive research. Instead, the engineers in charge of the plan considered the citizens' opinions and all other opinions different from theirs, as irrational and so they were to be excluded from the formulation of the problem. They had a very rational approach, it is a pity that it was oversimplified and did not consider most of the real problems affecting traffic policy. They identified, as a logical consequence of their approach, the single "correct" social behavior opposed to the "wrong" ones. The administration and the professionals working for the administration should somehow remove the wrong behavior. The ICE campaign was supposed to be the tool to pursue this goal set by the engineers.

In this circumstance, single engineers did not show arrogance at all, they were just naïve and absolutely unable to understand different aspects of rationality, not to speak of the possibility that even irrationality is part of the problem. As a group though, they had been put in the abusive position to influence all planning and decision-making processes. People's beliefs and behaviors, apparently inconsistent, may often hide rational interests that social research can identify. The engineers and the decision-makers did not expect any coercive action on behalf of the city government, but their idea was simply to apply their own approach and their plan. The engineers' problem was that they did not intend to waste a single minute on arguing their assumptions and on taking into consideration the possibility that others' opinions were acceptable and logical too. Not to speak about the possibility that, even if others' opinions were irrational, nonetheless it might have been necessary to deal with them anyway. The rigidity of the engineers' beliefs could have been either inspired by a conservative approach or by an environment-friendly attitude. In both cases, they refused the possibility and the necessity to let different ideas coexist and to let social change proceed in a non-linear direction.

From the point of view of communication policy, the problem can be synthesized in the following statement: the decision-makers and their engineer consultants did not consider that communication could have been bi-directional. They assumed only that they had the responsibility to inform the citizens, but they did not think that the citizens had something to communicate to them. This was neither conscious arrogance nor a desire to impose their opinion. PDMs and engineers intended to convince the people rather than impose a power that, in fact, they did not really have. They were uninterested in professionally comprehending peoples' preferences and in receiving information from the citizens because they did not know that this could have been an efficient action that would have led to better and quicker decisions. A one-way information flow from the administration to the citizens is a mistake from a communication policy point of view. This requires that citizens be taught how they should receive messages, condition them to accept them in their entirety, and finally approve – possibly unanimously – the behavior proposed by the PDM's. This approach does not work. There is no possibility that a local decision-maker could significantly modify people's behavior by manipulating perceptions even if helped by the best communicators. So all the money spent on this unattainable goal is wasted. A serious local communication plan for implementing a social change for a more responsive traffic and environmental policy should rely on the creation of sequences of constructive actions rather than cherry-picking a list of behaviors deemed to be "correct". Information about citizens' preferences is more useful than sending desperate messages such as "don't drive, take the bus" or the like. The voice of a local government can be almost completely drowned out by that of the national and global advertising and media.

One Single Thought for a Uniform Society

The data on citizens' behavior that I collected in the research mentioned above (Poli 1994b, 2006[6]) were not used by the administration to design a communication campaign. Those data were difficult to be used by professionals who did not have a competence in social science, namely by technicians and engineers in charge of the traffic plan. So the PDM hired, for a few bucks, a communication company to inform citizens about what they were supposed to do to fit in with the plan, which was confused with a kind of Procrustes' bed. Consultants in communication, hired by the PDM to inform the public of the traffic plan, sent a single message that was supposed to reach all citizens. They could not do more for the little money they received and for the lack of experience they had. Indeed, the consultants worked for a small company willing to do anything for a little contract. Engineers and PDMs were firmly convinced that it was not possible to do anything better than this. The communication company participated in one of the most egregious mistakes that administrations make. They sent a single message to all the citizenry assuming to speak to a hypothetical average person. This approach was exactly compatible with the idea that there was only one possible correct way to implement the engineers' traffic plan. Because all the other ways and options, in their opinion, were wrong *per se*, they were not interested in knowing *how* they were different – i.e. wrong. The goal was simply to redirect the "wrong" proposals onto the path of the one that was supposed to be the "right" one. The idea of informing the people with a good-for-all single message was quite arguable even in the mass society of the recent past. It is clearly unrealistic in the contemporary situation. An ICE plan makes no sense if it does not take into consideration contemporary social diversity. Therefore, a communication campaign, to be effective, should identify the diverse components of the people and classify them according to lifestyle, social and family values, income, education, age, gender, etc. All market surveys are developed in this way and recently even political parties have adopted marketing-like methods to determine what their constituencies really want. The same attention and the same skills are not used with the same attention for urban and traffic policies.

On the other hand, we can maintain that the lack of professionalism in information campaigns is one of the many hidden resources of democracy. As long as the sponsors of information campaigns confuse communication and education with manipulation and standardization of people's minds, their technical inefficiency is

[6] From this pamphlet (Poli 2006) I have drawn some ideas for this book.

welcome as a blessing in disguise, a revelation of the divine Grace that might not enlighten the sinners' minds but at least protect them from perpetrating evil.

From the Identification of the Citizens' Preferences to Scenario Building

Investigating citizens' preferences favors social change as it allows us to identify many diverse beliefs and attitudes that cannot be synthesized in the identification of an imaginary average citizen. The more we analyze the range of ideas spread throughout the population, the more we can design policies and make available choices that match with a multiform society. The preference analysis should permit the categorization of some homogeneous citizens groups. As in marketing, we can apply long-established statistical tools such as factor or cluster analyses. Once the citizens groups are classified, it is possible to design a policy coherent with needs and preferences of each group. When the mass society bias is not questioned, only one policy is devised to satisfy an average citizen. We can either consider society as the sum of individuals that merge into a single unit, or assume that society is an arena composed by groups that individuals join temporarily and among which they shift freely from one to another. It is possible to classify society into five to ten groups, identified for instance by statistical survey methods. In this way, we would be able to design traffic policies for each group. We could shift our attention from designing one single plan addressed to everyone, to focusing on the compatibility of diverse behaviors coexisting in the same urban and social system. Some of these behaviors and preferences will be really incompatible; others will coexist, some will compromise. This planning method is completely different from an approach that does not take into consideration the possibility of diverse contemporary solutions suitable to different groups of citizens.

For example, we can assume, as a second-level goal that we want to limit the use of cars so as to reduce emissions. If we adopt the average citizen abstraction, the plan will treat all citizens in the same way and the authorities will ask everyone for the same driving reduction. But the same driven-miles reduction does not necessarily imply the same sacrifice for everybody. Consequently, the PDM will not be able to influence some people who have good reasons to, or simply want to, drive more. Nor will the PDM be able to provide an infrastructural and organizational situation that will encourage people, who would be willing and able to drive less, to do so. This approach will prove:

(a) scarcely effective, because we can ask only minor reductions of everyone;
(b) very conservative, as it would not favor a re-thinking of the traffic policy and the realization of innovative infrastructure and mobility policies;
(c) possibly unjust as all citizens would receive equal treatment although they are in different situations;
(d) that injustice and ineffectiveness problems are likely to influence decision-making as they foster protest and opposition.

ICE and UMP Plans

The starting point of the UMP should be the ICE plan. Planning is a tool to favor change. Besides the construction of some infrastructure and a partial reorganization of mobility, the UMP should and could imply a change in citizens' behaviors, attitudes and habits. We wrongly assume that the revealed preferences – i.e. behaviors, attitudes, and habits – of citizens are consequences of their free choices, hence, preferences revealed in the past would be the same as what citizens aspire to for the future. This is obviously untrue and it is even less true in a rigid and complex society such as the contemporary city. The possibility to act in order to change some people's behaviors by applying the UMP should be the most important planners' concern. The plan's goals should be identified by an open ICE plan capable of letting citizens express their preferences – preferences which may be different from the ones that surveys revealed in the past.

Behavioral change is a hard task also because most people's habits are embedded in the urban social and physical structure. If we really want to adopt a rational approach, we should conclude that most of the reactions to planners and PDM proposals will be emotional. Many aspects of the planning process and most of the conflicts that arise during the planning process are hardly understood if approached in a mere rational way. To be really rational and practical, we must accept both irrationality in individual and collective human behavior, and the coexistence of different paradigms of rationality. For instance, we cannot circumscribe rationality to a simple cost/benefit/risk analysis because, for a long time, socio-economic and psychological research has proved that most people tend to assign a more significant value to their social position rather than to other considerations taken into account by economic calculations. In an affluent society, where "positional" goods – in Hirsch's terminology (1976) – have become more and more important for the achievement of individual happiness, the social aspects of traffic policies should receive more attention. These aspects are well known among sociologists and scholars who deal with the social change problem. In a traffic planning process they are not taken into consideration, both because it is difficult to immediately transform them into operative tools and/or decisions, and because they are foreign to the dominant technical/physical approach. *Per se*, an ICE plan does not rest on a single thought, rather it requires taking diversity into consideration.

If we adopt an ICE plan as the starting point of the UMP process we:

- should not have unrealistic expectations about a quick social change of citizens' behavior: the most we can expect is that we will start a process of change;
- although the UMP is directed to the entire population, we should be satisfied if we will be able to affect just a few sections of it;
- although we may forecast an optimal final situation, the plan should focus on change sequences.

Some people find it easier to live in a fantasy world and keep complaining about the mediocrity of a daily life that does not match their dreams. Others appreciate small, progressive, hidden advancements and they like to feel part of a long range project on which many cooperate.[7] The success of the plan should be assessed according to its capability to:

- (in general) speed up the social change processes already under way;
- (specifically) effectively apply some actions to immediately improve the current situation.

The ICE plan is not a short term one. It should be conceived as part of the UMP and should be an evolving apparatus that is constantly in use and continually developing, by no means a tool to use once, and then throw away. The ICE plan produces its effects in the long term. As an operative apparatus, the ICE plan is designed to constantly help in consensus building. It contains three types of information:

- Data (information),
- Communication tools (interaction),
- Consensus creating tools (decision-making methods and procedures).

A (quite possibly obvious) comment seems to be in order. We are accustomed to waiting for years before any given infrastructure is constructed, and throughout this time the promoters – and often citizens too – do not lose faith in the opportunity it will offer when eventually built. When we turn to social and communication plans we expect immediate results and, at the same time, such plans are often under-financed. If ICE plans were integrated into a UMP, authorities could take advantage of emergency situations that require (even if only temporary) traffic limitations by law, for instance because of poor air quality. If an ICE plan were in effect, the population might be more willing to accept some sort of traffic limitation, and the authorities could translate the emergency into an opportunity for people to learn new mobility behaviors.

The planning sequence could begin as follows[8]:

- A UMP is a structural plan that does not imply construction, but only organization and soft actions;
- If the UMP is structural, it implies that we are not expected to accept the current situation as carved in stone (i.e. growth trends, present mobility patterns, revealed preferences, etc.);
- Being a structural plan, a UMP is meant to produce changes: we assume that these changes should be behavioral, organizational, and also physical;
- We need to change the mobility structure without building any new physical structure.

[7] It would be interesting to discuss who is actually less presumptuous: technicians who dream perfection and believe in a single undisputed truth; or politicians/planners who aim at conflict solving in the frame of a low-key, but consensual project. From a certain point of view, the former are more modest because they fight for their own ideas and let the free-market competition of opinion choose the winner. The latter assume that they can understand others' ways of thinking, therefore placing themselves at a superior level.

[8] After having written a full book raising questions rather than giving answers, it would be odd if I now presented a solution, i.e. a UMP model. The example that I am going to present in the following sections, is nonetheless intended to expand the reasoning instead of giving a solution.

More Mobility, More Wealth

Mobility is not a goal: it is a means. The ambiguity springs from the broadly shared (but wrong) belief that growth in the movement of people will bring about better conditions. This belief has been implicit for a long time. It is responsible for having inhibited a different way of thinking. Thus, this dogma should be eradicated. It is not necessary to do a lot of research to prove that this equivalence is no longer true. Inquiries and studies may be nonetheless necessary to correctly measure how and how much the "more mobility/more wealth" dogma is false so as to more effectively reveal its fallacy to public opinion. Even though we cannot exclude – before research – the possibility that more mobility can actually bring more wealth, considering the disasters of the last 50 years, it would be wise at least to argue this opinion. The "burden of proof" ought to lie with those who have created the present mess and caused the never-ending protest of the people.

As a consequence of the "more mobility/more wealth" dogma, we have designed *mobility* plans instead of focusing on *accessibility*. All over the world, we design mobility plans, but there are few, if any, examples of accessibility plans. The obvious goal indeed is to have a better accessibility to what we need. Mobility is the means to access places. But we can also think in terms of making services available and reduce the mobility needed to get to them. If we adopted this approach, it is unlikely that in the short and middle run most of the decisions would be different from the usual ones, but in the long term this approach would help a radical change. Moreover, if this proposal is taken seriously into consideration, researchers will strive to elaborate consequent plans, tools, and techniques. The zero mobility growth hypothesis must be taken into consideration to (at least) reduce the mobility growth rates (if not yet the absolute values). Environmental concern requires this approach.

Before entering the project design phase of the UMP we should raise the following questions and try non-conventional answers:

- Why do we travel in urban areas?
- Why do we want to increase mobility supply?
- How would an increase in mobility benefit the community?

The priority goals of the UMP could be:

- Improve air quality,
- Make life more comfortable in the city,
- Reduce traffic accidents and improve safety,
- Reduce commuting,
- Increase walking and biking,
- Reduce energy consumption,
- Increase physical and virtual accessibility.

Of course, others might propose different objectives, but there is no reason not to begin by dealing with these options from the very beginning of the planning process.

We cannot help inferring that to pursue these goals, we need intellectual and social science tools different from the ones that are usually applied. For example, if we think that we should increase mobility as it is related to the growth of wealth, the containment of pollution would be a burden rather than the goal. But the traffic policy would be completely different if the goal is a drastic reduction of pollution and the burden is the conservation of a given level of accessibility through mobility. Similarly, we could establish that only 1,000 cars are allowed to circulate in a certain area. Then, the entire traffic policy should be designed to make life better under this limitation. This approach is the opposite of the usual one which presumes that: if 1,000 cars generate a traffic jam, we need to build facilities to host them ... and possibly even more. For decades, when construction of a mobility facility was finally approved, PDMs and engineers always agreed that it was an opportunity to make it bigger than the one that would have satisfied the current mobility demand. This was justified by the self-fulfilling prophecy that traffic would always grow and that it was "good". Thinking in this manner, PDMs intended to be forward-looking and progressive people.

From the Plan to the Projects

We need to adopt this communication approach to UMP in its entirety and from the beginning. Because the plan is based on the idea that: (a) all opinions are acceptable, (b) decisions should be made by searching for the broadest consensus possible, (c) citizens preferences and their requests should be satisfied as much as possible, hence, negotiation and mediation among conflicting values are welcome.

Such a new approach to the UMP does not have to achieve all its objectives immediately in order to prove itself. The most valuable function of the plan is to readdress the traffic policy in the long run. UMP:

- teaches citizens and PDMs to think in different ways;
- becomes an information and communication tool capable of facilitating decision-making;
- paves the way to new projects and investment opportunities.

In the middle run, this third function of a communication-based UMP can become the one that creates a link between the investors' community and the environmentalists. E.g., it is not necessary that, to reduce physical mobility, we immediately enforce laws that encourage tele-work or tele-education. The very idea that this is a possibility would stimulate investors to enter the market and take advantage of the opportunity offered because they now know they will be listened to. The result should be a hastening of the change processes. Eventually, environmental education is also an economic matter, at least for those who cannot help but think in materialist terms.

There are three ways to start an ICE plan:

- assume that authorities already know the situation so that we do not need more research and discussions to define goals at all levels;
- involve citizens by encouraging their questions and presenting them with scenarios; and possibly let them vote or at least express their opinions;
- let the city council or the elected representatives debate and vote.

It is not necessary to argue which one of these options is the absolute best. The most opportune strategy depends on the local circumstances. However, it is important to stress that only a legally and democratically correct procedure makes the plan legitimate and capable of functioning effectively. The planning phase, in which the goals are fixed, is too crucial to be kept implicit. Many of the conflicts that usually emerge in the course of the plan approval process, as well as during its implementation, are due to ambiguities about the real goals and about the way they have been chosen.

To design an ICE plan we need:

- a legitimate process both in the design and the approval phases;
- researches aimed at changing people's perceptions and at eradicating the usual analytic paradigms.

We do not need research that merely updates the standard information we have been collecting for decades. We need to shed light on phenomena that have never been examined. When Einstein elaborated new theories about light, scientists made hypotheses about the existence of pieces of matter that they had never seen and were not even sure really existed. Social scientists should adopt a similar approach and propose ideas and solutions that at present would seem unlikely to be accepted, but that well could be, once proposed and explained. Instead, we are completely in the hands of technocrats who can deftly elaborate existing data, but have neither the competence nor the attitude nor the scientific curiosity, to make new hypotheses. Because they skillfully handle operational tools, they succeed in imposing their own views that cannot help being conservative. Radical change is what we need, and now.

Chapter 11
Some Procedures and Some Content

Domi manere convenit felicibus

Apostolios[1]

Abstract An urban mobility plan's preliminary operations and researches are anything but routine as they determine the final recommendations and decisions. As a first step it is necessary to distinguish between dependent and independent variables. One of the basic questions is: (a) Is it better to analyze first the current situation and then think about how we can deal with the problems and modify the situation? (b) Or is it more convenient to identify the situation we want and then design a plan to make it happen? A further basic question is: When and if we find that groups of citizens have different preferences, should we provide a plan that is reasonably good for everyone or we better design a multiplicity of possible mobilities? Another crucial issue is how to deal with vested and diffused interests. Although different solutions can be equally legitimate and effective, I propose a radical approach based on reduction of mobility needs and on providing citizens with variegated solutions to their specific movement problems.

Keyword Revealed and expressed preferences • Vested and diffused interests • Social research • Variegated preferences • Decision-making • Methodology • Theoretical approach • Communication • Citizen involvement

Some UMP Procedures

In this chapter I will present a series of operations and preliminary researches necessary to design a UMP.

[1] "Those who are happy prefer staying at home".

C. Poli, *Mobility and Environment: Humanists versus Engineers in Urban Policy and Professional Education*, DOI 10.1007/978-94-007-1220-1_11,
© Springer Science+Business Media B.V. 2011

First, PDMs must be convinced that they need to intervene deeply in the process of modifying urban mobility. Customarily, the introductory research for a UMP focuses on traffic flows, parking demand and supply, mass transit users and so on. If we start by analyzing these data it implies that these are the only independent variables. Therefore, the plan will only gravitate to the goals of making traffic flow easier, give people more opportunity to move around, increase parking space and promote mass transit use. If this traditional research is the starting point of the planning process, the entire direction will be affected. Traditional traffic research, no matter how useless, is quite elaborate, thus it is expensive. A lot of money seems to be required to "precisely[2]" calculate traffic flows, the movement of cars and people, waiting times, and the like. As a practical matter, its cost is actually rather low because it is based on standard procedures applied in the same way everywhere by the engineering companies that are in charge of them. This research is sold at a high price because it is designed by highly qualified professionals (as the engineers presume they are) and because it is nonetheless only a small part of the total cost of the plan which in the long run is deemed to include the construction of heavy infrastructures. Because this basic analysis absorbs most of the resources appropriated to research, very little is left for social research. One might suspect that the real goal of the engineers' expensive research is to avoid social research that, if effectively produced, could challenge their beliefs.

If we still decide to do research on traffic flows, parking spaces, etc. but we choose to do it only after having developed a good knowledge of the social aspects of mobility and of the real needs of the people, the results would be simpler and cheaper. We would need to survey much less information. In fact, the plan would be based on the preferences the citizens have declared publicly and therefore that data would be collected in relation to the goals. Planners should ask this introductory question first: why should we intervene with traffic? This may seem an idle question, but a precise, meditated and shared answer to it would allow a coherent approach to the subsequent choices.

In fact, we should solve the following dilemma:

(a) Is it better to analyze first the current situation and then think about how we can deal with the problems and modify the situation?
(b) Or is it more convenient to identify the situation we want and then design a plan to make it happen?

The two approaches are logically alternative, but, during the planning process, it is likely that some overlapping may occur. If we adopt the first approach, it is unavoidable that planners, more or less consciously, will keep their preferences unspoken. Analyses are never completely neutral and independent from the researchers' goals. Therefore, the problem is not a question of neutrality or objectivity; rather it is that the goals are neither made clear nor perhaps even discussed. If we choose the

---

[2] "Precision" and data accountability are, by the way, a myth. In fact, data on traffic are typically random as they depend on thousands of unpredictable and unexpected human decisions.

second approach, the perception – also implicit – of the current situation affects the identification of goals. Even though neither one of the two approaches can be considered completely value-free, and the choice of goals will always be a political act instead of a merely technical one, the first approach is more conservative and the second more open to change.

If we chose option (a), it is not worth continuing reading this essay since we are clearly endorsing the second approach. If you think that option (b) is promising, then we have to go on solving new dilemmas arising in the course of the planning process.

As a first step for designing the UMP, the PDM, with the cooperation of professional planners, should define the planning process. This preliminary activity requires approximately a month or two if a well-trained expert is hired. Some PDMs consider that it is a waste of time to invest in this methodological work. It seems a theoretical activity. But theory is the most practical tool that humans have ever invented. Monkeys – who have other qualities humans lack – never apply theories, nor do people who despise philosophy. Hence, some preliminary theoretical thinking is likely to hasten the subsequent decision-making process. Indeed, a few weeks "wasted" to organize an efficient and legitimate planning process will save a lot of time later. It is interesting to note that two months to study the planning process and the decision-making system are, in reality, little time compared to the time that passes between the proposal of a public work and its actual construction. The ancient Romans used to warn: "*festina lente*" that translates as "hurry up, slowly". A PDM's refusal to discuss the planning method and the decision-making procedure is clearly instrumental in preserving the status quo.

The planning process presumes a correct, explicit and legitimate identification of the steps that a PDM should take to write the final draft of the UMP. These steps are approximately:

- Research on the planning method;
- List of necessary preliminary inquiries;
- List of research necessary during the plan design;
- Presentation of an ICE plan in which the communication phase will explicitly include a series of consultations with the citizenry;
- Design of voting and deliberative procedures to approve decisions in public bodies (councils) and/or in a people's referendum.

The planning process scheme must be public in order to favor citizens' participation in every phase of its design and implementation. For this reason it must be written and presented in a way that it is easy to read by everyone. To make the plan easily understood by non-specialists in urban and traffic policy may require hiring professional communicators and a few extra weeks. The planning scheme should also circulate among the people so that they are aware of what is going on. A detailed scheme should be exhibited in the City Hall and online. It is possible to adopt diverse planning procedures. Metaphorically, the planning process should be something in between a domino game and a puzzle composition. In the first we need to search for the most opportune connections; in the second we should try to fill all the empty spaces with the proper pieces.

The Citizens' Preferences

In social research we distinguish between revealed preferences and expressed preferences. If we analyze the citizens' present and past mobility behavior we refer to the revealed preferences. To a large extent, behavior is determined by previous decisions, by the availability of facilities, by services location, etc. For example, if mass transit service is poor, as a consequence I will commute by car. Therefore, a direct question such as "How do you go to work?" will necessarily bring the response that I drive. At the same time we can assess that the yearly private traffic growth in the last 5 years has been 5%. All this does not imply that trends are an inexorable destiny and that citizens really prefer to do what they have been doing so far. It is like asking paraplegic people if they have problems moving around instead of asking them what we can do to improve their lives. If we raise the first question, the answer cannot be but an obvious "yes". But this specific question, and the subsequent unavoidable answer, hint that we need to provide paraplegics more mobility opportunities. If we ask them how we can improve their living condition, the answer is more open.

The calculation of traffic flows and of people's inclination to move to areas far away from their jobs, is useful to explain what has already happened, but it does not necessarily give any information about the future. We better wonder whether what happened in the past will continue in the future, and whether it is really what people want. The analysis of the revealed preferences does not say anything about how the citizens imagine their future. Focusing too much on the past favors inertia, inhibits creativity and posits a conservative bias on the planning process. Repeating and diffusing information about the past, and assuming that the trend is immutable, works as a self-fulfilling prophecy. Researches on revealed preferences have little to do with a planning and social change policy. The future cannot be imagined by projecting the past into the future. Although some data collection on past and present situations, and even on possible trends, is still useful, the second, more important, step should be research designed to allow citizens to express their opinions about the future. I do not dare to deny that, to a large extent, the questions raised to identify the citizens' expressed preferences will be biased as well. However, researchers can do their best to carry on serious research that limits value judgments and/or makes them explicit. This researchers' endeavor is not aimed at getting closer to a supposed truth. It is rather a procedural device to increase knowledge. The questions raised will help to give information about new possibilities offered for solving mobility problems and can evaluate how real is the possibility, for example, of using a bicycle or a bus instead of driving and in what conditions; or the possibility of looking for a job closer to home, or of working from home (tele-work).

The research on expressed preferences must be very serious and sophisticated otherwise it is not worth doing. Besides inquiring about the preferences, attitudes, and behaviors of the citizens, the research should be structured in order to achieve two more fundamental goals:

- Increasing the information available to the citizens;
- Facilitating the decision-making process by linking the research to legitimate voting or choosing systems.

The Mobilities Plans

Families have different mobility patterns in respect to singles. Young people have different needs compared to the elderly. Within the same age groups, classes, genders, and social status situations, people's behaviors and attitudes differ widely. In contemporary urban society, different cultures and life styles live side by side. There are also gender differences: research has shown that women and men have dramatically diverse attitudes toward driving, transportation, and environmental protection. Differences are no longer a consequence of traditional working habits – women at home, men at work – but are related to more profound and complex reasons (genetic or cultural?). Moreover, social and gender groups cannot be catalogued any longer merely by markers such as: sex, gender, income, or education. We need to identify people's attitudes that go across many traditional analytical classifications. Accordingly, we should speak of a mobilit*ies* plan rather than a mobility plan. A good UMP should offer options to the diverse urban cultures. The preliminary research on the social aspects of traffic policy should allow us to:

- Collect information scientifically on the social universe;
- Regroup citizens according to their lifestyles, beliefs, behaviors, attitudes toward change, and so on;
- Involve the population in a participatory activity to decisions from the beginning of the planning process;
- Spread information about new tools, general options and solutions;
- Shift more easily from social research to a better-informed decision-making process.

Diffused and Vested Interests

Social research would identify the opinions of non-organized citizens; politicians should give them a voice. However, people are also members of associations – vested interests – which influence the planning process by having a more direct contact with the PDM and mass-media, and thus by controlling votes (and money) in the elections. It is very difficult to get rid of the influence of vested interests and, to a certain extent, it is also ethical[3] and convenient to take them into consideration. Social research is meant also to reestablish a balance between the excessive weight of vested interests and the weaker voice of citizens when they act in a non-organized way. If this happens, politicians may win new constituencies formed around the new citizen's preferences and goals that researchers have helped to identify.

In this essay, we have constantly stressed the necessity of favoring the communication of the people's needs and opinions to the PDM. If the decision-makers do not take public opinion into consideration, protests are likely to occur. It is expedient

[3]In a competing society, which is far from being an ideal free-market of politics, we may need organizations to defend and affirm some threatened rights in some circumstances.

for PDMs to avoid protest despite their belief that they will be able to prevail over citizens' opposition. Conflict is time consuming, therefore the time dedicated to litigation should be listed in the cost of every project. Politicians often underestimate this aspect. It is a long-standing tradition that has become part of the anthropology of political behavior: politicians actually enjoy fighting, because it is a way of imposing their ideas, interests, opinions. Also, their prestige among their supporters (including the powerful vested interests) depends on their ability to win over citizens' opposition through negotiation, as well as by showing determination. In reality, this attitude is unjust and time consuming, that is to say, completely inefficient. If, instead, the PDM promotes a planning process which includes a decision-making system able to ease deliberations, they will have exercised their power in a relevant and positive manner. They will still have many opportunities to enjoy fighting.[4]

Three Approaches to Traffic Policy

We can choose among three approaches to traffic policy:

- Traditional,
- Moderate reform,
- Revolution.

Whichever we choose, we must keep in mind that the goal of a good plan should be coordination and enhancement rather than removal of possible options.

In the traditional approach to traffic policy, planners and decision-makers rely mostly on revealed preferences. This implies that current trends are followed and reinforced. The rationale of this approach is that choices are made following the preferences revealed by the market. If there is more traffic, we must build more roads; if large shopping centers are built and are successful and the people move to the suburbs, this is what we should accept. PDMs are supposed to design plans that matche what citizens have been shown to prefer. Putting it simply, if roads are filled with cars, it means that we have to build more roads to respond to the demand. This traditional approach would be a practical and safe response. It is not necessary to change anything, and we trust that capital and technologies will always be available to build the necessary infrastructures. Environmental problems are treated in the same way. If air pollution grows, we foster research for lower impact engines and fuels. If we ask about the environmental impact of the new road construction, the response is that better circulation eliminates bumper-to-bumper highly polluting traffic, and that good design can even improve the landscape.

[4]It is interesting to notice how politicians adopt an aggressive, war-like, and thoroughly "macho" language, but in reality the crucial behavioral pattern of decision-making actors is negotiation and mediation. Should we conclude that political discourse is the encounter of a traditional male language with a female behavior?

Obviously, today, traditional traffic planners are not as naive as to focus only upon road construction. Although this remains the core of their traffic plan philosophy and finance, they also look at public transportation policies for which governments are now making conspicuous funds available. They also consider the pedestrian areas and bike routes that are requested more and more often by citizens and environmental activists whose demands are proving worth responding to in order to get consensus. However, bike routes and pedestrian areas are provided with the perspective of responding to an existing demand rather than with the idea of increasing what is offered in order to promote a possible unsatisfied, hitherto hidden, demand. The basic assumption of this approach is that an increase of mobility automatically increases the collective well being. All the obstacles to more mobility are considered burdens: this means that all the problems which occur because of this increase in mobility are dealt with by applying technical solutions incrementally rather than by understanding and interpreting their real nature.

This approach has many flaws, but also some factual strong points, not necessarily connected to the political support received from engineers and constructors lobbies. Many citizens approve of this approach because they feel safe when simple, well-known and effective solutions are offered. Moreover, the damage caused by this approach is evenly allocated. It is true that some citizens will be harmed by the construction of a highway just in front of their house and that the value of their property will decrease as will their quality of life. However, in the course of time, this always happens to "others": the constructors and PDMs seldom face mass protest. A road construction may directly distress 1,000 inhabitants in a metropolitan area of one million, most of whom may benefit more or less directly from the new construction or may be not affected by it directly. The citizens damaged by the new construction "in the interest of the community" are left by themselves to protest, and they are easily defeated. Such a decision-making process can be judged abusive if we think of it as imposing a series of small damages disguised as solutions. But if we consider this approach the only possible solution, its ability to avoid citizens' aggregation is very valuable. To contrast this way of thinking, we can rely only on a strong ethical commitment of people who believe that more construction is simply wrong for a series of reasons concerning environmental protection issues. In the case of wishing to accomplish large construction such as a beltway or a light rail network, the problem involves the entire city and all its voters. Here, the entire decision-making process becomes a general political issue. Consequently, it is possible to have both tough opposition and deep-seated support to the single major project. Nonetheless, the discussion would consider only the single project rather than the overall traffic policy.

There are two sound objections to critics of the traditional approach. First, while it is true that the preferences revealed in the past are conservative and they do not say anything about what people would like to have in the future, nonetheless, it can be assumed that in the short and middle run these preferences are relatively stable. Second – and this is the most reasonable objection – is it likely that, if citizens are asked to express their preferences for the future, they will propose something that has not yet been tested? Is it not possible that by asking people what they would prefer, we are in fact imposing our own approach, since all the research is shaped on the bias of the interviewers? Finally, the citizens might express preferences for solutions that are not technically feasible.

The second possible approach suggests that we try to invert the growing trend in car use. This approach, like the first, is based on revealed preferences and does not attempt to dispute them. It also deals with the traffic problem by starting from a solution that is immediately available and already considered sufficient to solve the problem, namely, that it is enough to increase public transportation. This dogma is deeply rooted in the thinking of most environmentalists. The supporters of this position tend to focus on this solution and to neglect other aspects of the traffic congestion and mobility problem. Of course, they also are more open than conservatives to the construction of pedestrian areas and bike routes. However, it is equally apparent that the investment in construction and operation of mass transit facilities overwhelms the trifles left to all other projects. Hence, the plans inspired by this (apparently) progressive approach are still a list of projects and solutions, although they are oriented towards public transportation.

This approach is different from the conservative approach as it welcomes solutions acceptable to people who are concerned with environmental problems. The solutions proposed by those who adopt this approach address the preferences of a significant group of citizens and, at the same time, they are welcomed also by vested interests. This approach is favorable to the usual heavy investment in public works. Investments are partly shifted from road construction to mass transit facilities and infrastructures. Moreover, investments in public transportation do not completely substitute for investments in road construction[5] because an increase of mobility supply may favor an overall increase of mobility that in the middle term will require also more roads and car use. Investment in public transportation has rarely shifted drivers to buses or rails. If at some time it did happen, it was a minor amount. The success of a new mass-transit facility is measured in terms of an absolute increase of public transportation users, rather than as a shift from cars to buses or rails.

This second option matches the sustainable development (or, as we should say, sustainable growth) idea. Underneath its presumed and flaunted environmentalism, all that the sustainable-growth-environmentalists do is to convince industrialists to invest in more environment-friendly "solutions". Thus, in the attempt to involve traditional industry in environmental projects, the sustainable growth environmentalists have become lost in convenient negotiations and have forgotten the real goals. The sustainable-growth-environmentalists' efforts are aimed at being financially "sustained" and are tolerated by the political industrial establishments of which they are progressively becoming a part.

[5] In many urban areas, there is literally no more space to build new roads. Therefore, the only possibility for public investments is in public transportation facilities. The option to shift conspicuous investments from new construction to intense maintenance and improvements is, instead, generally neglected.

The third approach is the radical one that I have been proposing all through this book. It is based on the idea that a further increase in people's mobility is socially harmful and is not convenient even in terms of business as usual. Admittedly, this statement should be proved. Economists could elaborate a model to examine the effects of a shift from traditional constructions and mobility related investments towards innovative projects aimed at reducing mobility and environmental impact. Most economists' research is now focused in proving the positive effects of investments in transportation for the general economy. Certainly investments in transportation have cumulative consequences in economic growth: the question is if different investments can have better consequences, possibly also on traditional economic growth.[6]

A decrease in physical mobility is not considered an option either by politicians or by citizens – and even less by technicians and professionals who have an interest in a constant growth – but it has been considered an option in environmentalist literature, i.e. by marginalized writers and scholars. However, some traffic reduction projects are becoming popular and have been applied here and there (Engwicht 1989; Gottlieb 2007). The problem is making them become the mainstream of traffic policy instead of just small appendages to regular solutions. Alternative projects are really effective only if they: (a) are part of a general plan; (b) are consistent with the plan's overall goals; (c) are not neutralized by other decisions pulling the system in the opposite direction.

Point (c) is the most serious. In fact, in current times it is very easy to be open to everything, so that establishing priority is difficult. Thus, we cannot expect to design a plan by using only one of the three approaches, that is, we cannot expect to be 100% radical and forget about people who would prefer more conventional solutions. I argue for a procedure that will lead to a radical change, but I do not reject temporary traditional solutions approved in order to satisfy the current demand or to cope with some emergency. At the moment, the real practical advantage of this third approach is that it allows us to identify a radically new approach, no matter how difficult it may be to apply it immediately.

A Mobility Reduction Plan

The basic concept of the radical approach to the mobility plan is as follows: the plan's goal is to reduce the need to move while keeping the income level, and the quality of the services available, constant. Moreover, with this approach, mobility is not considered *a priori* as the one possible solution to other problems: a traditional example being that, if some people lack goods and services close to the area where

[6] Economists may study models to prove the consequence of a shift of investments from one industry to the other. E.g., Leontief's input-output analysis (1966) was widely used in the 1970s and 1980s to re-address economic structure and is still often applied (Dietzenbacher and Lahr 2004).

they live, facilities need to be built to provide citizens access to services in wider and wider areas. This is not a solution at all to discard and it must be evaluated comparatively with the option of having services or goods available as close as possible. As a well-known traffic environmentalist planner argues: “Just as a bread wrapper is a waste we produce in getting access to bread, our car trips are a waste we produce in getting access to ‘products’ – like coffee with a friend, our work, a hair-cut or the groceries” (Engwicht 1989). Traffic policy should be approached with the goal of identifying techniques and tools capable of reducing the production of waste. Consequently, the planner adopting this approach should have these priorities:

- How can we substitute intensive energy consuming movements with less energy consuming ones, beginning with walking and proceeding to biking and presumably to public transportation and private car use?
- How can we eliminate non-necessary movements by making individual mobility more rational?
- How can we reduce the length of each trip?
- How can we organize the urban settlement to reduce the environmental impact?
- How can we organize and design services, economy, technologies and social life so as to reduce the mobility needs?

These questions should not be raised only in the technical and decision-making milieu – i.e. among the specialists working at the plan – but they should become a shared way of thinking among the people. I have raised questions, instead of having proposed answers. If the UMP were approached by trying to offer an answer to these questions, all the results could be different. Because with the previous two approaches we have never resolved any real mobility problem and have instead constantly worsened the situation, any wise person would think it would be worth just trying.

The Funniest Phase of the Planning Process

If these are the questions, the answers (i.e. solutions) must come as a consequence from them. Some solutions already exist, others must be worked out. At this point, and not before, technicians are allowed to enter the stage. Civil and transport engineers will now be welcome. They have been rather mistreated in this essay – but not for their own responsibility, rather for the role that they have been assigned and have often been abused for.

This third phase of the planning process is the easiest and the funniest. In this phase, citizens and decision-makers develop and learn about new solutions. The financing of many projects and research can be suggested here. While worrying about methods, justice, and social change can generate anxiety and discomfort in people who are really concerned with the common good, this part of the plan implementation is more relaxing and can be used as a therapy after the torment of the methodological phase. In the following chapter, I will present a number of these proposals. Some of them have already been applied and implemented; others are just hints offered to inspire thinking and new ideas. Before listing these projects and these possibilities, I want to make clear that this is not what this book is about. I intend to offer a different way of thinking about traffic policy and do not want to propose just new solutions. So what follows is useful only if (a) the projects listed will be the core of the traffic policy and not just an addition to the basic road and infrastructure building; (b) if these projects are considered specific initiatives which would be meaningless if not sustained by the philosophy described above.

Chapter 12
What to Do?

I sometimes think drivers don't know what grass is, or flowers,
because they never see them slowly …
My uncle drove slowly on a highway once … they jailed him
for two days.
… Have you seen the two hundred-foot-long billboards in the
country beyond town?
Did you know that once billboards were only twenty feet long?
But cars started rushing by so quickly
they had to stretch the advertising out so it would last.

Ray Bradbury, Fahrenheit 451°

Abstract Most citizens/consumers currently give priority to issues such as (a) saving energy, (b) reducing pollution which brings serious effects on human health, (c) contrasting the decay of both natural and constructed environments. A series of projects is described in this chapter. They are introduced by a reasoning about the political content of the solutions proposed. Some of them have already been applied, some are being designed, other are ideas to be developed.

Keywords Traffic and environment • Citizens preferences • New professions • Political support of new economy • Barriers to change • Retail policy • Urban design • Housing policy • Public transportation • Telecommunications

The Political Issue

In affluent countries, road construction encounters more and more opposition. Eventually, most of the new construction proposals are realized since the construction industry exploits the atmosphere of emergency and panic they themselves are able to create. High-impact public transportation facilities are also controversial, although to

C. Poli, *Mobility and Environment: Humanists versus Engineers in Urban Policy and Professional Education*, DOI 10.1007/978-94-007-1220-1_12,
© Springer Science+Business Media B.V. 2011

a lesser extent. These days, governments are more inclined to finance a mobility policy and a traffic infrastructure if it can also be justified by the necessity to reduce environmental impact. An infrastructure is less likely to win popular consensus and be financed if it is proposed merely in the name of economic growth.

At the top of the policy priority list, a growing number of citizens/consumers place issues such as (a) saving energy, (b) reducing pollution which brings serious effects on human health, (c) contrasting the decay of both natural and constructed environments. Governments should create the market conditions to allow the economy to respond to citizens' new needs. There is plenty of evidence in social research that a growing number of citizens are currently dissatisfied with urban livability. They would be ready to buy alternative solutions to their mobility patterns such as trading fewer transportation facilities with more livable areas which do not require driving many miles in order to go shopping, to school, to work and/or to use a wide range of services. Unfortunately, there is little and (therefore) expensive supply for this demand, which, on the other hand, is not yet well defined. If this new potential alternative demand really exists, it might be satisfied by the market.[1] If the market does not work, because too many subsidies are assigned to the traditional transportation and infrastructure constructions, governments should recreate the conditions to make it work again and fulfill the existing demand. If we want a dynamic economy, we must favor innovative sectors instead of going on sustaining a mature industry such as construction.

From a procedural point of view, we can assume that there are both political and economic stakeholders who have an interest in change. There are those who (a) politically support decision-making methods based on citizen participation, and (b) technically propose the adoption of new solutions. These parties might find valuable allies and powerful sponsors in those investors who have an interest in a shift from the traditional appropriation of public money from physical infrastructure to developing economic sectors. Among these parties, we have both investors and professionals who look forward to applying new skills in newly created job markets.

Politically, all these stakeholders are on the verge of initiating a change process. They will succeed if they would support the grassroots movements and the neighborhood advocacy groups who give voice to the new demands and highlight new needs. Citizens who already live in well established high-density communities are more keen to propose innovative and environment-friendly solution to traffic congestion and pollution than people who move frequently and/or live in single family units with little contact with others, like in suburbs. Other citizens will remain more favorable to road constructions and to the standard solutions. In fact, lack of community life and interpersonal communication makes it difficult to explain more complex projects and people tend to rely on simplified solutions. To a certain extent, this process of change is already taking place.

[1] Besides the usual market of goods and services, we mean also the so-called political market, i.e. the existence of a political demand by the citizens that politicians are interested in fulfilling. This approach to politics was inaugurated by the celebrated Anthony Downs' "*An Economic Theory of Democracy*" (1957) and still inspires an abundant literature. Although in general I am not a follower of an economic approach to politics, I admit that in this case it helps us to understand the current situation.

It proceeds slowly because the people involved often do not have a clear perception of the political implications and the strategies necessary to implement it. It also happens because the potential technicians and investors in new approaches to traffic problems – although innovative from the technological point of view – are often politically conservative. Namely, they (think to) belong to the same social class of the traditional construction industry who are, at present, their real competitors. This alliance between innovative economics – to which new professionals and soft-industries belong, instead of the "let's be practical"/material constructors – and the environmentalist movements has not yet happened because we lack a widespread political environmentalist culture. The environmentalist movements remain loaded with technical scientific approaches that prove ineffective in promoting political change.

The Traffic Revolution

Major barriers to change are:

- the structure of public expenditure;
- the power of the companies working in the transportation sector;
- the competence of the professionals who are in charge of traffic and transportation plans;
- consumers' preferences and their difficulty in knowing about and adopting new possible behaviors.

Prima facie, we propose three main actions in the traffic and transportation planning process:

- citizen participation that includes:
  - identification of the citizens' behaviors and preferences;
  - public information and communication about new possible solutions to the traffic and transportation problem;
  - identification of patterns of behavioral change;
  - future visioning;
  - legitimization of information and decision-making processes;
- design of soft solutions such as:
  - tele-work;
  - logistics;
  - telecommunication and information technology in a broad sense;
  - economic and fiscal tools;
  - a different legal framework;
  - land use;
- higher education (university) programs that educate new professionals to deal with the traffic problem in a creative way and with a strong focus on social and philosophical environmental issues

Given that we consider traffic revolution a comprehensive political issue based on theory and political philosophy, we are entitled to suggest some practical strategies and tools available to readdress traffic policy. All the policies listed below are likely to be more cost efficient in traffic reduction than the building of new infrastructures. Some of them have already been applied, some are still projects waiting to be implemented, others are just ideas.

Commuting Reduction

A flexible housing policy – based on both renting and ownership – can make possible a closer proximity of jobs and housing. This would reduce commuting. Laws and policies – e.g. incentives – should be addressed mainly to the middle class, which includes the majority of population. An overall reduced mobility (commuting, in this case) is in the interest of the most vulnerable sections of the population, such as asthmatic children and elderly people who risk pollution-related diseases. Middle classes are in the position to reduce their mobility since they can afford to spend money to improve their quality of life. However, they must be encouraged to act so because nowadays, there is no market where they can easily "buy" the environment-friendly lifestyle they desire. At present, a shared political discourse is not yet available to deal with these kinds of justice problems that have traditionally been focused on income and class inequalities. This proposal is listed first because it represents a complete inversion of the usual policies, a U-turn, to be precise. In the last 50 years, decision-makers have often tried to solve everyone's problems by increasing mobility.

This mobility-reduction-addressed housing-project originates in the following question: is it more cost effective to build a new throughway (or a light rail) to increase the overall transportation supply by 5%, or to invest the same money to put residences and jobs closer in order to reduce commuting by 5%? If every morning 30,000 people go from A to B, and we have another 30,000 who go from B to A for the same reasons, is it possible that, by implementing a reorganization project based on information, tele-communication and economic incentives, 5% of the commuters may find a job closer to their house (or a house closer to their job) in order to travel less? If we were speaking about a 20% or 30% or 40% quota, the proposal would not have much credibility. But 5% is a realistic goal. To be achieved, we do not even have to invest the tens or hundreds of millions necessary for the construction of roads and rails, as well as for their maintenance, nor waste millions in the environmental (ignored) costs their construction and operation imply. In Europe, a significant part of traffic is generated by school and university students and by some huge bureaucratic organizations whose reorganization has never been considered *vis-à-vis* the traffic problem. The personnel of large factories and bureaucracies (including schools and hospitals) are heavily unionized. There are many activist organizations, and sometimes they are politically influential because of the specificity of their requests. The students, too, have their organizations to advance their requests to the authorities. However, most requests focus on better transportation rather than on other possible solutions, leading to an inefficiency of jobs and services location. Research and communication should be advanced to inform people on alternative means to solve their problems.

Every 5 or 10 years, municipalities should encourage a reshuffling of jobs in order to bring jobs closer to homes. There are several possible means to pursue this goal. Helping match demand and supply on a limited territorial basis would help a significant quota of the people to find a job closer to their residence.

City and Neighborhood Design

Urban design and regulation should ease the creation of urban villages and communities. The success of the Starbucks-like coffee shops proves that many contemporary urbanites (most of them singles) look for places of aggregation and consider the coffee shop around the corner an extra living room of their small (or smaller) apartment. This need of physical proximity is an asset on which traffic policies may be able to rely on. This current movement-averse people attitude should be taken into serious consideration.

Accessory Uses of Cars

Related to urban design there are other interesting projects about traffic and car use reduction that originate in a simple consideration: in traffic policies, the car is considered only as a means of transportation, but it is also and often used for other side-purposes. For example, many use their cars to store items they do not or cannot carry – like shopping bags, exercise and sport equipment, extra clothes and shoes. A car is also a benefit when it rains, and sometimes we find ourselves driving a few 100 ft merely in order not to get wet. Cars are a relaxing place to sit in, a wandering living-room equipped with stereo, air conditioning, heat, and comfortable seats. Sometimes being in the car is really a good experience: who hasn't had great conversations while driving through distressed urban areas oblivious of the external world while outside it was raining hard, or was icy cold, or so humid that it would have hampered your breathing if you were outside? Thus the question becomes: How can we (partly) substitute these uses of polluting cars? In order to avoid people using cars for secondary, non-transportation highly-polluting uses, we should endow the city with a series of minor structures including covered paths, luggage deposits, toilets, comfortable rest and meeting places which can compete with cars. If we design cities' public spaces more properly and make them livable, people would reduce their car use and would appreciate walking, biking or just sitting in fine cafes.

How difficult is to make this kind of city design really effective? It is very difficult because we are not used to investing money in research, education and development of the skills required. How effective would these projects be in reducing traffic? Very little, for sure! Let's say something like 1% or 2% each? Hopefully, in the course of time a little more, once the skills have developed. But this is not the question worth raising. We should rather consider the following three matters:

- How much do all these projects cost compared to other projects implemented to reduce traffic? It is likely they cost almost nothing if compared to a new light-rail, or a new road. The cost effectiveness of the dollar invested in reducing traffic is likely much higher
- If these projects all together succeed in reducing traffic by 5% overall, how much will congestion be reduced? It will be reduced by more than the 5% of traffic reduction because it will eliminate cumulative effects
- What are the cumulative effects and the virtuous sequences begun by these projects? They would be in place of the traditional "macho" and muscular way of thinking, which sees the only solution in attacking the congestion problem with heavy, material and expensive constructions that set in motion vicious circles instead.

More Local Information Reduces Traffic

A community newsletter and/or a web site with proper local information about all the services offered in the neighborhood can reduce traffic more than a new road. Local information, including business advertising, festivals, group meetings, sport events, and so on, helps to congregate a community and fulfills more needs locally. Once started, this approach generates cumulative effects and starts virtuous circles. It is likely to promote local investments.

The General Contest of the Urban Design and Traffic Reduction Policies

Favoring bicycle use is, of course, a positive step toward reducing car circulation. However, local authorities, besides providing bike routes, bike renting, and the proper design of urban and suburban streets, need also to address apparently indirect issues like safety and security, facilities for a more widespread use of the bikes (dressing rooms at work places, convenient storage, bike-parking, rentals and repairing services). The general context in which bikes are supposed to be used is more important than merely providing a basic bike route. Even though there are plenty of implemented projects that favor biking in the cities, I do not know any place where the government has intervened with a significant investment comparable to one usually assigned to road or transit construction. Even when bike routes are abundantly provided, most of the funding regards construction and little or nothing operation.

The Neighborhood Designed by the Citizens

In some cities, municipalities promote the so-called "*charette*" design for the neighborhood. People cooperate in designing the area in which they live. Many possible experiments can stem from this approach and creative solutions can be tested.

Favor Short Distance Public Transportation

Public transportation policies should focus on the neighborhood level more than on the metropolitan region. For instance, a transportation service supplied by 5–10 vans (12–15 seats each) available on call and serving an area of approximately 10 ha (approximately 2,500 acres) and 20,000 people can fulfill most of the local mobility needs without fostering long distance commuting. On long distances, public transportation is not so effective in reducing car use for two main reasons: (a) users need to drive to get to the bus or railway stations; (b) being too far away from their own car reduces flexibility and the possibility to use the car for uses other than personal mobility.

Retail Policies

Other projects, that would significantly help to reduce traffic, regard retail distribution. Local governments should invest in research, design and money to favor the establishment of neighborhood shops able to provide services required by the local population. For example: cafes, organized spaces, civility places (comfortable meeting points attended by polite people), or home services delivering food and cooked meals. The question is, how much would it cost to develop marketing research and, subsequently, a project? How much could it reduce traffic? We do not know, but "no investigation, no right to speak" (Mao Dze Dong). The lobbies of large scale distribution will not be very happy with this policy. They may even boycott research about what proves helpful to favor neighborhood retail distribution and related services. Indeed, incentives to small businesses in retail distribution should be deducted from the money given to finance road construction which is an implicit incentive to large scale wholesale distribution companies. Free market supporters should not be scandalized if the local or central government helps small business start-ups with some minor tax relief. They should instead protest against the heavy support given to large scale distribution via infrastructure building. Such a policy is very revolutionary because it endangers big companies and big politics. It is even more "dangerous" because it would challenge some social class rooted biases. Small business people would abandon the conservative parties dominated by real capitalists and join the supporters of mass democracy. We should never forget that large retail organizations and the constructors of huge shopping malls are intertwined with, and heavily impact, the financial market.

Shop in the Neighborhood Card

Years ago, as a member of an Italian middle-size city administration, I tried to implement a project called "shop in the neighborhood card". The local government provided know-how (the idea and a feasibility project) and involved the associations of local retailers in the project. The associations were interested since they were supported by many small retailers who were fighting daily against large wholesale companies endangering their family-owned shops. On behalf of the local government I negotiated some incentives to be exchanged with some civic services directly or indirectly provided by the retailers in the neighborhood. There was a possible market in most of the neighborhoods into which the city was organized. The growing number of elderly people, who cannot have easy access to suburban stores, was one good public policy reason to promote the project. The project was quite comprehensive, but the most interesting aspect was the "shop in the neighborhood card": a discount card given to those who lived in the neighborhood and decided to choose the local shops instead of driving far away to the shopping center.

Re-colonization of the Suburbs

In the suburbs – where local shops and everyday services are not available within a reasonable distance – municipalities should buy (or favor the purchase by small business investors) a few scattered lots and promote the establishment of "villages" by encouraging the settlement of new business serving local needs and operating as social aggregation hubs. Something in this line was tried in Brisbane, Australia in the nineties.

Local Marketing Research

We need research about what is available and what is missing in the neighborhoods so as to provide possible investors with a proper market knowledge and favor investments targeted to reduce mobility. Incentives can be associated to this research to assist new and inexperienced possible investor entrepreneurs to prepare business plans to locate loans and funds.

Telecommunication and Information Technology Applications to Traffic Reduction

Telecommunication and information technology is now normally used to patrol and control access to some traffic limited areas or to collect tolls. It can also be applied to regulate the average speed on roads and highway in order to avoid traffic blockage. It would have a threefold positive effect: eliminating part of the pollution by removing the stops and goes; providing a better comfort to travelers; informing drivers about the trip duration. We all know that it is better to know that we will arrive in a longer, but guaranteed time, than being unsure when we will be able to make it. To implement this approach, a cruise control device should be distributed and made mandatory, especially in Europe where it is not yet extensively adopted, so to eliminate jams and stop and go traffic with positive effects also on safety.[2] Tele-work projects have been popular for at least 20 years and now they have been effectively implemented by many companies. The problem with tele-work projects is that most of them were not developed with the goal of reducing traffic or environmental impact. Most tele-work projects have been conceived in order to save office space or make life easier for women with children. These projects never took off effectively because no laws and effective policies have been applied.

If we approach tele-work from a traffic-reduction point of view, we should not assume that a growth of tele-employment would reduce total commuting in a metropolitan area by more than 10%. In the short term, the effects of a tele-work policy cannot be dramatic. However, if convincingly pursued, a tele-work policy can prime social change sequences in the direction of keeping the people more connected with the neighborhoods in which they live. A population that works more at home and commutes less would have profound effects on the neighborhood's social and economic organization.

We would need at least 5 years before seeing effects of a tele-work policy on commuting and social re-organization. The promoters of projects like tele-work and house/job re-location are always urged to prove their immediate effectiveness. People lack faith in the potential of tele-work and relocation to solve their problems, and are not eager to change their lifestyles. The communicators' goal should be to illustrate the change sequences and thus create "time": having a clear outlook of the final expected and welcome situation helps people to bear uneasiness during the time needed to adapt to the new situations. Since these are the major problems for implementing a tele-work/traffic reducing project, we cannot leave the implementation of such a project in the hands of engineers, even tele-communication engineers who are, generally speaking, more flexible than civil and transportation ones. In the past, most of these plans were designed by engineers hired by private enterprises eager to sell their technology rather than to solve social and environmental problems. Other professionals (psychologists, urban planners, political scientists, economists, sociologists, whatever, etc.) have played a marginal role in tele-work project-design. Decision-makers – often seduced by the enterprises' promoters – prefer to commission a project to technicians who promise immediate solutions rather than consider a real policy whose effects will show in the course of time. To a partial justification of decision-makers, I admit that there are few professionals in social sciences effectively trained to design these projects. I myself, when I had local government responsibilities, could not help hiring engineers for a telecommunication project for which

[2] In a best-selling book, Tom Vanderbilt (2008) has broadly argued about driving styles that may be changed. Also traffic and mobility psychology scholars have accumulated a lot of information about driving habits, but it has not been applied as much as desirable.

sociologists would have been a better fit. It was not possible for me to hire them because: (a) the council, which was skeptical about my approach, did not trust sociologists and thought it would have wasted money; (b) I was not able to defend my project because I myself did not trust the skill of the sociologists available.

Between Public and Private Transportation: Solidarity Transportation

Five years ago I developed a project in order to extend the idea of Public Transportation to private cars. This proposal is extensively described in the following chapter.

Chapter 13
Between Private and Public: Mutual Transportation

Do not go where the path may lead; go instead where there is no path and leave a trail

Ralph W. Emerson

Abstract The discussion about transportation has been polarized on public vs. private transportation, e.g. bus vs. cars. In this chapter I present an intermediate solution, i.e. using private cars as public transportation. We have called the model "mutual transportation"; developed in 2005, it has been presented in several conferences and proposed to a number of local government and traffic authorities. It is now applied in some cities and metropolitan areas. Besides the technical solutions, which are based on telecommunication technology, the novelty of the system proposed is that it has been devised and designed by humanists rather than by engineers, who have been hired later to solve some technical problems identified by social scientists.

Keywords Private vs public transportation • Mutual transportation • Park & Ride • Moscow (Russia) • Traffic reduction policies • Restricted transit areas • Car-pool • Traffic authorities • Security • Telecommunications

Introduction

In this final chapter I present a project that was developed by social scientists and designed to limit urban traffic. I am aware that it is "a solution" and so it partly contradicts my reiterated claim for elaborating open plans instead of designing operative tools. However, it is useful to submit it as an example of a project whose origins are in social science. It was possible to have and develop this idea because I focused on the social aspects of traffic. If I had been thinking only about how to build new roads and bridges, I might never have had this idea. I developed the idea

C. Poli, *Mobility and Environment: Humanists versus Engineers in Urban Policy and Professional Education*, DOI 10.1007/978-94-007-1220-1_13,
© Springer Science+Business Media B.V. 2011

and the project, starting from a social analysis of people's preferences and behaviors. The project was developed as part of students' training at the "Laboratory of Environmental Communication" at the University of Bergamo, Italy. I presented it in several conferences and to a number of city governments in Europe and in the US. Several versions of it are now available with different names and there are also some actual applications. Notwithstanding its easiness to apply and its potentiality, its diffusion is proceeding slower than expected because of a lack of interest and competence in this approach to traffic policy.

The idea and the project are very simple. Nonetheless, it is difficult for the project to be accepted because it is unfamiliar to the managers who deal with traffic problems in the transportation departments of local governments. Even though it might have a discernable effect on traffic reduction, city administrators do not take it into consideration as seriously as it would be a road construction project. I already mentioned that scarce consideration is given to these soft projects because of the dramatically reduced amount of money they move. Also, there are no organized influential guilds to sponsor this "solution". As a consequence, citizens remain uninformed and are, therefore, reluctant to embrace the new system.

Arguing the Separateness Between Public and Private Transportation

We need to fight a bias concerning the rigid ideological separation between public and private transportation. Public transportation is traditionally provided by large mass transit infrastructures. The progress in public transportation has been always perceived as:

- an extension of the networks,
- the augmented capacity of the transportation means,
- an increase of transportation means speed,
- an extension of the areas served, i.e. spreading the transportation network into broader and broader suburban areas.

More recently, with the hypertrophic extension of the suburbs it has become less and less feasible for public transportation to serve all residential areas. Thus, new transportation policies have favored the construction of longer bus or rail lines able to reach farther developed areas instead of intensifying the service offered in smaller areas. This illogical settlement and transportation pattern is the origin of the Park&Ride "solution". Mentally lazy environmentalists welcome this solution as a progress toward a more widespread use of public transportation but they do not question the real problem: Park&Ride organization increases both the use of the public transportation and the use of cars by allowing people to live farther away from jobs and services.

Only recently and still in a partial way, transportation planners have taken into consideration issues such as comfort and time, and itinerary flexibility. Improvements have been marginal because of the overall rigidity of the transportation structure. Moreover, small and middle sized cities adopted the public transportation technologies and systems that had been designed for large metropolises. A more careful analysis would have persuaded transportation planners that these systems were inefficient. Instead of solving the problems, the adoption of big cities' transportation systems have brought into small cities the problems of the metropolises. In small and middle-sized cities, it is worth adopting a new idea of public transportation that does not require heavy constructions and infrastructures. The project I will describe is initially addressed to approximately 10% of city travelers but this percentage is likely to grow in the future. We should abandon the prejudice that public transportation requires a massive organization and enormous (mainly public) capital investments. I assume that public and private transportation are not necessarily as separate as they are generally considered. Private citizens could voluntarily – or possibly be encouraged with some incentive – make their cars available to satisfy part of the mobility demand. In contemporary society, new technologies can favor a more efficient use of private capital – i.e. cars – that is currently underemployed.

The basic rationale of the "mutual transportation project", is a better utilization of existing capital, namely private cars. The political and social by-product of such a project is the creation of a collaborative attitude among travelers. This project does not reduce personal mobility, rather it makes it easier. This should be considered a minor pitfall of the project if we were to adopt dogmatically the idea that mobility should not be encouraged and instead should be reduced.

A Russian Origin

The idea for the project came about when I lived in Moscow. My idea is based on the so- called "gypsy car service". It is used largely in Moscow and other Russian towns by both local residents and visitors. In Moscow, it is very simple to get a ride in an unofficial taxi market. Muscovites simply wave at any car as if the cars were all cabs on duty. It is a kind of hitch-hiking, except that you pay for the ride after having bar gained with the driver. Foreigners are amazed to see how many cars stop to offer rides as soon as someone raises their hands. Generally, only older cars stop, but in some cases you may get a ride in a fancy brand new car driven by a chauffeur who wants to make some extra money while traveling without the owner. You usually do not wait for more than a minute. Frequently you receive more than one offer. As you negotiate the itinerary and the price with the driver of the first car that stopped, others drivers may stop and wait patiently in line, hoping your negotiations were unsuccessful. The presence of other cars waiting to offer a ride favors the rider in the negotiation. Some drivers cruise around the city in order to find customers and consider this as a real job or at least a second job. Still, most of them are ordinary travelers who want to make some extra money or even like to have some company while commuting.

Foreigners are surprised by the confidence people show in accepting a ride from a stranger. Of course, they are also surprised that the drivers let strangers into their cars. Moscow is a huge metropolis that nobody considers a safe or relaxed place. Would something similar be conceivable in New York City? Or in any other big western city? Probably not. Then, the relevant question is: why and how does this surprising system work? Besides many local cultural reasons, the system in Moscow is mainly based on trust and need. The system developed after the fall of Communism and the paramount increase in private car ownership. Most of the drivers and the passengers do not take any real precaution to prevent being robbed or threatened. They are simply not so risk-averse as most paranoid westerners. However, in some cases, mostly in the peripheries and at night, I noticed that as soon as I agreed on the location where the driver would drop me off, they would call someone on their cell phone. My Russian is not good enough to understand precisely what drivers were saying, but by the few words I could understand and by the tone of their voices, I realized that they were calling for security reasons. My partner, an American teacher of Russian language, told me that drivers just told someone where they were going. This simple information was nothing but a smart message to the passengers: drivers let the passengers know that somebody knew whom they were with and where they were going. If passengers misbehaved they were easily traceable. Also passengers were likely to call someone for the same reasons. We could analyze and argue a lot about the anthropology of the system, namely the reasons and the conditions that make the system work. Nonetheless it is much more interesting to describe: (1) why this system can inspire traffic policies in other cities and (2) how the conditions that allow the system to work can be created.

The Characteristics of Mutual Transportation

Let us begin by describing the idea and the project. It is a simple and very effective system to limit car circulation in urban areas. The basic idea is to encourage higher car occupancy. It is an evolution of the largely adopted, but not very successful, car-pool and/or car-sharing systems. The design is very simple from a technical point of view. It requires a project designed by a sociologist with the cooperation of an engineer specialized in telecommunication and a traffic (civil) engineer. The whole project, however, requires good organization, economic and social research, and includes a professional communication campaign. Moreover, being new and innovative, the designers must be prepared to adjust the project during the design phase and implementation.

Expected Effects on Traffic Policy

The system is effective because it:

- increases car occupancy,
- reduces the number of vehicles circulating,
- does not substantially affect (reduce) the overall mobility supply,

- enhances travelers' satisfaction,
- can be integrated with other traffic policies,
- reduces parking demand,
- reduces public transit overcrowding during rush hours,
- promotes social relations.

What Can Be Accomplished

It has the same effects as car-pooling:

- allows cars with more than two riders to use reserved lanes and to have access to restricted transit areas,
- allows cars with more than two riders to circulate when traffic limitation measures (usually caused by air pollution) are applied,
- makes it easier for public authorities to enforce stricter traffic limitation measures that can be applied more extensively in time and space,
- facilitates streamlining toll application and access to city centers.

The Difference from Standard Car-Pooling

The difference with car-pooling is the way in which cars are filled, i.e. how travelers get rides and drivers offer rides. The main problem with car pooling is its lack of flexibility, which mainly affects the passengers. Passengers and drivers must agree before leaving home and use the telephone or internet to match their scheduled rides. In order to favor higher car occupancy, the traffic authority designs the following system:

- Drivers must be encouraged to offer rides spontaneously to whomever asks for them[1]
- Commuters and travelers must be encouraged to ask for rides instead of using their own cars.[2]

The potential passenger puts their thumb out to hitch-hike. To allow for traffic safety, the Traffic Authority can mildly regulate the system without losing flexibility, by providing pick-up points or areas, for example. Of course, drivers and passengers will also be provided with a system to plan rides in advance and use internet and/or cell phones to search for a ride.

---

[1] Some people are compulsory drivers; others cannot help using their cars for some good reasons. Consequently, it is likely that there will always be a ride supply.

[2] Many people might dislike driving, want to save money, and prefer not to take a bus or public transportation which moreover might not be available in time and space, etc.

How Can It Be Convenient for the Driver?

Those who decide to drive their cars must find it valuable to offer a ride to other travelers. Car drivers with more than one (or two[3]) riders would:

- save on road tolls – which can be more broadly applied,
- save on parking fees,
- be able to use reserved lanes and have access to restricted transit areas, which can also be multiplied and enlarged,
- drive when traffic limitations are enforced, as often happens mainly due to air pollution,
- benefit from an economic incentive apt to cover at least part of car operating costs.

Moreover drivers might like to share some company and meet new people during their trip.

Convenience for Occasional Travelers and/or Regular Commuters

The system must be also convenient for the passenger. As a matter of fact, if a good balance between ride demand and supply is created, the system is very flexible and efficient as possible passengers:

- may wave at any car and get a ride whenever and wherever needed in the city,
- may like to share some company and meet new people during their trips,
- do not have to wait for a bus and/or can try to get a ride while waiting for them,
- do not have parking problems.

Advantages for the Traffic Authority

The traffic authority can easily create good value without a significant investment. As a matter of fact, an investment is justified in the interest of:

- having less traffic,
- generating less pollution,
- saving in road and parking construction and maintenance,
- saving in (often subsidized) public transportation
- creating a cooperative environment.

[3] Some research and specific further design is necessary to make the system fit into local situations. E.g., having two riders (besides the driver) in the car might offer extra incentives, even though it might make the system more complex.

Offering this opportunity, the Traffic Authority becomes more entitled to multiply access and turnpikes tolls and in general increase or establish road prices for cars carrying the sole driver.[4]

Possible Problems and Solutions

Competition with cabs: When I was interviewed about this project by a newspaper reporter he immediately concluded that the cab drivers' organizations would burn my car right away. He was wrong though, because the system is not an alternative to taxi service. Of course, some attention should be directed towards making sure it does not interfere with taxi service. We should avoid a Moscow-like situation where the system is absolutely illegal and has virtually swept away any legal taxi service. The solution is that the system must be free for the passengers and based on cooperation and some possible public incentives.

Competition with Public Transportation: The CEO of a major mass transit company was frightened that this system would discourage customers from using public transportation and that they would choose to accept private rides instead. This objection is quite weak. First, in Europe only 35–40% of the budget of public transportation companies comes from fares or tickets. The remaining is subsidized by the governments. Second, the project works on marginal numbers of the total transportation demand and supply. Also we propose it as part of the overall public transportation system. To have access to it, customers – both car drivers and passengers – would utilize (and exhibit) a mass transit company membership card. In this way the public transportation companies would be involved as well. In this case some investment would be moved from financing construction and purchasing transportation means to the implementation of the project, in other words from a physical capital-intensive service to a soft service. The public transportation company would maintain control and could be interested in promoting the project as part of its mission. In this way, we are in a position to involve all the traveling citizens in a public transportation activity with the organization of the mass transit companies. To resolve security problems we can use as identification the transport authority ID that usually entitles commuters to special fares as it occurs in most European cities.

Security

How secure can a system be in which you let a stranger into your car or, vice versa, you step into a car driven by a stranger? This seems the major problem for many drivers and passengers. The system can easily be made more secure if both drivers

[4] Prices for all cars might be established, taking into consideration the possibility that the passengers and the drivers are encouraged to share road costs.

and passengers use a viable identification network that, as proposed above, can be the Transportation Authority membership card. *Prima facie*, the steps to join and organize the system might be as follows:

- people are invited to register at the Traffic Authority office;
- the office registers the identity of the member (both car drivers and possible passengers), and provides the driver with a bumper stick (or similar identification) to be applied to the car;
- the same person applies as a driver, as a passenger or (more advantageously) as both;
- the office also provides an ID to be displayed in the car and/or to be exhibited under request;
- the passenger who enters the car will feel more secure knowing the driver's name and registration number;
- moreover they can contact (or in case send an automatic Text Message to) a call center, established by the Traffic Authority, letting it know that they are riding with a registered person whose code number they are able to provide;
- the driver is supposed to do the same, i.e. sending a Text Message with the passenger's code number;
- passengers and drivers itinerary can be traced by GPS;
- a call center may be established to facilitate the encounters.

More Possible Problems and Solutions

Regarding matching Demand and Supply, and Economic Incentives, informing the Traffic Authority's call center has a twofold goal: (a) security and (b) data collection to monitor the system. The system must be accurately monitored in order to make ride demand and supply adequately match. We assume that the main goal of the project is to reduce the number of circulating vehicles by increasing car occupancy. In a first phase, it is possible that fewer people are available to offer a ride versus the ones who want to take it. Passengers asking for rides will likely be the ones who are already familiar with using public transportation.

Beyond the general efficiency of the system, extra incentives for the users can be:

- free bus tickets,
- free parking bonuses,
- discounts,
- other.

They can be awarded according to the data collected by the automatic call center monitoring system. Privacy considerations are not an issue because the system is based on voluntary acceptance of the rules. The ride is supposed to be offered without any charge for the passenger. The Traffic Authority is supposed to create the conditions to make the ride offer convenient to the driver. In a second phase, depending on the results of the first tests, passengers might be asked to contribute to the drivers' expenses. The system will work at its best when participants are open to offering and asking for rides, exempt from bureaucracy, registration, incentives, etc.

Conclusion

This project has to be considered as an idea to break new ground. It needs to be adapted to local situations. Its success depends on social and physical factors, such as the ability of people to relate with each other and the city layout. Besides the practicality of the system, whose main goal is traffic reduction, it also implies a concern for a more open society and for people willing to cooperate and interact.

The system fits best in middle sized cities whose metropolitan area populations range from 300,000 to 600,000 inhabitants. We also proposed it in some smaller municipalities located in city suburbs and for organizations such as Universities. At the University of Bergamo, as an assignment of my course for the Laboratory of Environmental Communication, the students developed and proposed this system for reducing car traffic in the campus area. Most of the campus in Bergamo is located in the medieval center of the city and pollution in all the metropolitan area is often beyond the alarm or risk limits fixed by the EU. In Bergamo, as in all Italian cities, students commute from their home towns using either cars or public transit. This initiative of the students and the implications of their project was welcomed by most of the students and faculty, since they considered it a viable means both for traveling and for meeting new friends. However, as mentioned in Chap. 8, in the same weeks we were proposing this opportunity to cooperate in reducing traffic, the University President (Rettore) urged the municipality to provide more parking places near the campus. Obviously, this policy would encourage more students and faculty to commute by car.

I report this episode to describe how difficult it is, in the real world, to propose tools that are different from the usual ones. The system fitted in perfectly with student behavior, attitudes and social milieu. Students, being young and actively involved in their education, are likely to be more open to new actions and are particularly inclined to meet new people and to share amenities. The organization of Italian universities with their preference for having students commute rather than building student housing also favors the implementation of the project in that milieu. The system also fits with the metropolitan geography and the physical structure of the city and of the campus. Nonetheless, at decision-making time, the students' proposal to adopt this system was completely overlooked and the usual pattern won the priority.

The first experiment was conducted in Trento (Italy). As a test we chose a major road to access the city from the nearby villages. Trento is an Alpine town of 100,000 inhabitants. The metropolitan area houses approximately the same number of people. The city is located in a valley surrounded by mountain villages and towns from which people commute mainly by car. We began the test between Mesiano – a peripheral village now surrounded by suburban low-density settlements – and Piazza Venezia, a large square downtown that attracts a lot of traffic. The municipality of Trento designed a good, though traditional, traffic plan and has built more parking lots than most Italian cities. However, mobility still relies a lot on private cars.

We decided to begin the test by asking a group of volunteers to hitchhike along the road selected. In the beginning we did not provide any information to the possible drivers interested in offering rides. We employed ten volunteers (five men and five women) who hitchhiked in different places along the road and the city streets. Even without any advertising about the initiative, the average waiting times decreased considerably in weeks. We calculated the choices to offer and give rides according to gender, look, age, ethnicity and attitude. Women were more likely to be offered rides; men were more eager to offer, as it had been expected. From this first phase we inferred that feeling relaxed about offering rides is, to a certain extent, a matter of becoming accustomed to it. The more people are seen hitch hiking, the more confidence there will be in both asking and offering rides. To report some data: the first week the Average Waiting Time (AWT) was 22 min for men and 6 min for women. After 10 weeks the AWT shortened to 11 min for men and 2.5 min for women. Besides gender, we collected more detailed data about the characteristics of the drivers and the passengers such as how many people were already in the car, the relation between weather and availability to offer rides, and so on. We also calculated the places where it was easier to get a ride, a security perception indicator, and so on. We made a detailed survey of possible users. In the second phase, by interviewing a sample of possible users and asking them if they were available to give and/or take rides, we spread information on how to use the system and explained some of its characteristics. In the second phase we will advertise the entire project except the part related to the fees. We will employ both volunteer riders and drivers. The cars will be labeled with a visible bumper sticker or little flag in order to make them very visible and encourage other drivers to learn about the project.

These tests are not an introduction to the implementation of the project. We will need:

- to know how people behave in order to design the details of the project;
- to advertise the project: namely let the people get used to it gradually.

As expected, the municipality, although more open than others to the project, is not going to appropriate a considerable amount of resources. We calculated that with approximately €100,000 – i.e. the cost of half a city bus – we could manage the system for 2/3 years without considering the income obtained by the possible application of a fee. The investment would include a cell phone calling center, the possible fee application costs, call tracing systems, advertising and information, monitoring, general organization and continuous planning of the free/subsidized parking and bus tickets award linked to the system.

In the case of 1,000 rides per day, we could get rid of 1,000 cars from the city and would no longer need 1,000 new parking places. 1,000 rides is a minimal forecast in a metropolitan area with 200,000 inhabitants, with more than two million trips per day. If you compare these figures with the possible reduction of car users shifting to buses, the gain is evident considering the improvement in the comfort of the users. However, administrators know that they can easily find €200,000 in local or state finances for buying a new bus or opening a new (short) road, but they consider it almost impossible to fund this cheap initiative with adequate resources. The good news is that the system can also be applied by investing less money and on a voluntary basis.

Technical operative problems can be difficult – but also fun – to resolve. Among others, they include legal issues concerning insurance, the design of pick-up sites, the call center structure, the financial plans concerning the incentives. What really matters, in the spirit of this book, is that the system can be implemented and managed only if:

- somebody gets the idea, believes, promotes and sustains it;
- if the project is designed by professionals in different fields;
- if money is shifted from other items

Therefore this implies a threat to the civil traffic engineers' business as usual. This is the only reason why this project is not so often implemented or it is not as well known as it should be and could be.

Chapter 14
Conclusion

Man can survive a month without food,
a week without water,
but only twenty minutes without a justification

Jay Fisher

There is a time for building and a time for thinking about what is worth being built. Gone are the days when so-called facts were more important than arguing and elaborating ideas. Since when the Sustainable Development compromise has turned out to be a fraud – see Chap. 2 – we need to contrast the incrementalist approach of doing things at any cost, with the culture of thinking, and perhaps also that of planning. In contemporary urban mobility and environmental policy we should abstain from making things, at least big things such as physical mobility infrastructures. Any physical construction we avoid building is more environment friendly than even the most ecological structure we actually fabricate.

I began this essay by declaring that my goal was humanitarian, not intellectual, political rather than scientific. I did not aim at uttering any ultimate scientific truth or at proving any sound statement. I may also admit that not all the arguments proposed herein are fully elaborated. As a matter of fact, the object of this book is to present an unusual perspective in approaching urban mobility policies in order to reason on contemporary environmental crisis. This book is a research project rather than a completed report. Its possible value lies in the research it will be able to trigger on the grounds of the arguments developed and the clues suggested in each chapter.

The Slovenian philosopher Slavoj Žižek – though a (peculiar) Marxist himself – claims that the time has come to reverse Marx's oft-(over)quoted 11th Thesis on Feuerbach that goes as follows: "Philosophers have hitherto only interpreted the world in various ways; the point is to change it." Žižek argues that "the first task today is precisely *not* to succumb to the temptation to act, to intervene directly and change things , … , but to question the hegemonic ideological coordinates."

C. Poli, *Mobility and Environment: Humanists versus Engineers in Urban Policy and Professional Education*, DOI 10.1007/978-94-007-1220-1_14,
© Springer Science+Business Media B.V. 2011

Likewise, in Chap. 2, I claimed that we need a new environmentalist ideology whose possible development was curbed by the affirmation of the Sustainable Development pseudo-ideology. Žižek maintains that "those who really want 'to do something to help people' get involved in (undoubtedly honorable) exploits like *Médicins sans Frontières*, Greenpeace, feminist and anti-racist campaigns, which are (…) even supported by the media even if they seemingly enter the economic territory (…) – they are tolerated and supported as long as they do not get too close to a certain limit" (2006: 238).

People who feel content, and even consider themselves as morally commendable because of their minute contribution to the solving of minor problems by doing things, should instead be censured since they cooperate in keeping a wrong system in operation. They are even more reprehensible because they waste time that could be better employed "*expostulat(*ing) *why day is day, night night, and time is time*" which does not always mean wasting "*night, day, and time*".[1]

Because educational institutions are the ideal breeding grounds for idle argument, it is there that a real change might begin if innovative research was promoted. Unfortunately, today the citadels of knowledge have grown old and resilient to innovation. Therefore we have little hope. But hope, to be true hope, must be little, otherwise it gets confused with possibility and certainty. In his *Prison Notebooks*, Antonio Gramsci wrote that the "old is dying and the new cannot yet be born. In the interim, a great variety of morbid symptoms appear". We should all work to give the new a chance to be born by elaborating new discourses that, sooner or later, will aggregate in useful ideologies capable of addressing action.

"*If discoursing on a difficult problem were like carrying weights, when many horses can carry more sacks of grain than a single horse, I would agree that many discourses would do more than a single one; but discoursing is like coursing, not like carrying, and one Barbary courser can go faster than hundred Frieslands*".[2] This book may not be a very fast Barbary horse, but it is surely faster than any Friesland. Many Barbary horses are what we need right now: no matter if we do not yet know where they are heading to, but we need them to run fast and light in order to pave the way to that environmentalist revolution that we humans need to save ourselves and the planet.

[1] Shakespeare, *Hamlet*, act II, ii, 89–90.

[2] Galileo, *Il Saggiatore*, as translated in Italo Calvino (1988).

Bibliography

Agamben G (2005) State of Exception. University of Chicago Press, Chicago
Arendt H (1998) The Human Condition. University of Chicago Press, Chicago, first published in 1958
Badinter E (2010) Le conflit. La femme et la mère. Flammarion, Paris
Baumann Z (1993) Postmodern Ethics. Blackwell, Cambridge
Beck U (1997) The Reinvention of Politics: Rethinking Modernity in the Global Social Order. Polity Press, Cambridge
Benzaquen A S (1998) Thought and Utopia in the writings of Adorno, Horkheimer and Benjamin. Utopian Studies 9(2):149–161
Breese G (ed) (1969) The City in Newly Developing Countries. Prentice Hall, Trenton
Butler J (1990) Gender Trouble. Feminism and the Subversion of Identity. Routledge, New York
Calvino I (1988) Six Memos for the Next Millennium. Harvard University Press, Cambridge
Capra F (2007) The Science of Leonardo. Inside the Mind of the Great Genius of the Renaissance. Doubleday, New York
Carson R (1962) Silent Spring. Houghton Mifflin, Boston
Castells M (1989) The Informational City: Economic Restructuring and Urban Development. Blackwell, Oxford/Cambridge
Castells M (1996) The Rise of the Network Society (The Information Age: Economy, Society and Culture), vol 1. Blackwell, Cambridge/Oxford (second edition, 2000)
Castells M (1997) The Power of Identity (The Information Age: Economy, Society and Culture), vol 2. Blackwell, Cambridge/Oxford (second edition, 2004)
Castells M (1998) End of Millennium (The Information Age: Economy, Society and Culture), vol 3. Blackwell, Cambridge/Oxford (second edition, 2000)
Clark G (1983) Interregional Migration, National Policy, and Social Justice. Rowman & Allanheld, Totowa
Crozier M (1964) The Bureaucratic Phenomenon. University of Chicago Press, Chicago
Crozier M (1995) La crise de l'intelligence, Essai sur l'impuissance des élites à se reformer. Intereditions, Paris
Crozier M, Friedberg E (1980) Actors and Systems: The Politics of Collective Actions. University of Chicago Press, Chicago
Currie G, Worrall J (eds) (1980) The Methodology of Scientific Research Programmes: Philosophical Papers, by Imre Lakatos. Cambridge University Press, Cambridge
Daly H (1977) Steady-State Economics. The Economics of Biophysical Equilibrium and Moral Growth. Freeman & Co, New York
Daly H, Cobb J (1989) For the Common Good. Beacon Press, Boston

C. Poli, *Mobility and Environment: Humanists versus Engineers in Urban Policy and Professional Education*, DOI 10.1007/978-94-007-1220-1,

© Springer Science+Business Media B.V. 2011

Davies K (1969) The urbanization of the human population. In: Breese G (ed) The City in Newly Developing Countries. Prentice Hall, Trenton

Davies K (1972) World Urbanization Process: 1950–1970. University of California Press, Berkeley

Dietzenbacher E, Lahr M (eds) (2004) Wassily Leontief and Input-Output Economics. Cambridge University Press, New York

Downs A (1957) An Economic Theory of Democracy. Harper, New York

Eckersley R (1992) Environmentalism and Political Theory. Toward an Ecocentric Approach. UCL Press, New York/London

Engwicht D (1989) Traffic Calming: The Solution to Route 20 and a New Vision for Brisbane. CART, Ashgrove

Ferré F, Hartel P (eds) (1994) Ethics and Environmental Policy. The University of Georgia Press, Athens

Feyerabend P (1993) Against Method. New Left Books, London

Florida R (2002) The Rise of the Creative Class. Basic Books, New York

Fox W (1990) Toward a Transpersonal Ecology: Developing New Foundations for Environmentalism. Shambhala, Boston

Fox W (ed) (2000) Ethics and the Built Environment. Routledge, London

Friedman T (April 15, 2007) The power of green. New York Magazine

Georgescu-Roegen N (1971) The Entropy Law and the Economic Process. Harvard University Press, Cambridge

Gibson-Graham J K (2006) A Postcapitalist Politics. University of Minnesota Press, Minneapolis

Giddens A (1991) Modernity and Self Identity. Polity Press, Cambridge

Goleman D (1995) Emotional Intelligence. Bantam Books, New York

Gore A (1992) Earth in the Balance: Ecology and the Human Spirit. Houghton Mifflin, Boston

Gottlieb R (2007) Reinventing Los Angeles: Nature and Community in the Global City. MIT University Press, Cambridge

Habermas J, Luhmann N (1971) Theorie der Gesellschaft oder Sozialtechnologie – Was leistet die Systemforschung? Suhrkamp, Frankfurt

Hardin G (December 13, 1968) The tragedy of the commons. Science 162(3859):1243–1248

Hardt M, Negri A (2004) Multitude: War and Democracy in the Age of Empire. Penguin Press, New York

Hirsch F (1976) The Social Limits to Growth. Routledge & Kegan Paul, London

Hirschman A O (1963) Journeys toward Progress: Studies of Economic Policy-Making in Latin America. Twentieth Century Fund, New York

Hirschman A O (1970) Exit, Voice and Loyalty. Harvard University Press, Cambridge

Hirschman A O (1977) The Passions and the Interests. Princeton University Press, Princeton

Hirschman A O (1982) Shifting Involvements: Private Interest and Public Action. Princeton University Press, Princeton

Holden E, Norland I T (2005) Three challenges for the compact city as a sustainable urban form: household consumption of energy and transport in eight residential areas in the Greater Oslo Region. Urban Studies 42:2145

Jamieson D (1988) Sustainability and beyond. Ecological Economics 24(2):183–192, now in Jamieson (2002)

Jamieson D (2002) Morality's Progress. Clarendon Press, Oxford

Jamieson D (2008) Ethics and the Environment. An Introduction. Cambridge University Press, Cambridge

Jamieson D (2010) The question of the environment. In: Rodotà S, Tallacchini M C (eds) Trattato di Biodiritto, vol 1. Dott. A. Giuffrè Editore, Milano

Jasanoff S (1995) Science at the Bar: Law Science and Technology in America. Harvard University Press, Cambridge

Jasanoff S (2010) Testing time for climate science. Science 328(5979):695–696

Jonas H (1979) Das Prinzip Verantwortung: Versuch einer Ethik für die technologische Zivilisation. Insel-Verlag, Frankfurt am Main

Kozlowski J, Hughes J T (February 1967) Urban threshold theory and analysis. Journal of the Town Planning Institute 53(2):55–60
Latour B (1987) Science in Action: How to Follow Scientists and Engineers through Society. Harvard University Press, Cambridge
Latour B (2004) Politics of Nature: How to Bring Sciences into Democracy. Harvard University Press, Cambridge
Latour B (2005) Reassembling the Social: An Introduction to Actor-Network-Theory. Oxford University Press, Oxford
Leontief W (1966–1986) Input-Output Economics. Oxford University Press, New York
Leopold A (1949) A Sand County Almanac. Oxford University Press, Oxford
Lilienthal D E (1944) Democracy on the March. Harper & Row Publishers, Inc, New York
Marsh I (ed) (1991) The Oxford Book of Travel Stories. Longman Chesire, Melbourne
Meadow D H et al (eds) (1972) Limits to Growth. Universe Books, New York
Moos M et al (June 2006) Does design matter? The ecological footprint as a planning tool at the local level. Journal of Urban Design 11(2):195–224
Morris D (1967–1969) The Naked Ape: A Zoologist's Study of the Human Animal. Jonathan Cape, London
Mouffe C (2005) On the Political. Routledge, London
Myrdal G (1968) An Asian Drama: An Inquiry into the Poverty of Nations. Twentieth Century Fund, Inc., New York
Myrdal G (1975) The equality issue in world development. In: Nobel Prize Lecture to the Memory of Alfred Nobel. Nobel Foundation, Stockholm
Nagel J H (1987) Participation. Prentice-Hall Inc, Englewood Cliffs
National Research Council (1983) Risk Assessment in the Federal Government: Managing the Process. National Academy Press, Washington, DC
Nelson R (1977) The Moon and the Ghetto: An Essay in Public Policy Analysis. Norton & Company, New York
Neuman M (2005) The compact city fallacy. Journal of Planning Education and Research 25:11
Nijkamp P (2006) Book review. Planning Theory 5:271–283
Northern Virginia Transportation Authority (2006) Trans-Action 2030: Transportation for Today and Tomorrow. NVTA, Fairfax. http://www.thenovaauthority.org/transpublications.html
Plumwood V (1991) Nature, self and gender. feminism, environmental philosophy, and the critique of rationalism. Hypatia 6(1):3–27
Poli C (1994a) The political consequences of an environmental question. In: Ferré F, Hartel P (eds) Ethics and Environmental Policy. Theory Meets Practice. University of Georgia Press, Athens
Poli C (1994b) La città sicura. Rischi ambientali e politiche di intervento urbano a difesa della salute. Guerini Studio, Milano
Poli C (1994c) Epistemic issues for methodological and procedural reforms. Project evaluation and environmental risk. In: Proceedings of the 4th Conference on Food, Agriculture, and the Environment, University of Minnesota, Wayzata
Poli C (ed) (1994d) Etica ambientale. Teoria e Pratica. Guerini. Milano
Poli C (2006) Rivoluzione Traffico. Meno mobilità più comunicazione. Robin Edizioni, Roma
Poli C (2009) Città Flessibili. Una rivoluzione per il governo urbano. INSTAR Libri, Torino
Poli C (2010) Sustainable development: from fallacy to fraud. Human Geography: A New Radical Journal 3(2) pp. 63–82
Poli C, Timmerman P (eds) (1992) L'etica nelle politiche ambientali. Fondazione Lanza Gregoriana, Padova
Poli C, Timmerman P (eds) (1993) Ethics and Environmental Policies. United Nation Human Dimension of Global Change Programme and Fondazione Lanza, Toronto/Padua
Ray P H, Anderson S R (2000) The Cultural Creatives. Harmony Books, New York
Richmond J (2005) Transport of Delight. The Mythical Conception of Rail Transit in Los Angeles. The University of Akron Press, Akron
Rodotà S, Tallacchini M C (eds) (2010) Trattato di Biodiritto. Giuffrè, Milano

Rostow W (1960) The Stages of Economic Growth: A Non-Communist Manifesto. Cambridge University Press, Cambridge

Royal Academy of Engineering (2008) Philosophy of Engineering, Second International Workshop. http://www.raeng.org.uk/societygov/philosophyofeng/pdf/abstract_papers.pdf

Salvemini G (1939) Historian and Scientist: An Essay on the Nature of History and the Social Sciences. Harvard University Press, Cambridge

Shephard P, McKinley D (eds) (1969) The Subversive Science. Houghton Mifflin Co, Boston

Shrader-Frechette K S (1985) Science Policy, Ethics, and Economic Methodology. D. Reidel Publishing Co, Dordrecht

Shrader-Frechette K S (1991) Risk and Rationality. Philosophical Foundation for Populist Reforms. University of California Press, Berkeley

Shrader-Frechette K S, McCoy E (1993) Method in Ecology. Cambridge University Press, Cambridge

Singer P (1991) Environmental values. In: Marsh I (ed) The Oxford Book of Travel Stories. Longman Chesire, Melbourne

Stern P C, Fineberg H V (eds) (1996) Understanding Risk: Informing Decisions in a Democratic Society. National Academy Press, Washington, DC

Stevens M L (2000) Bioethics in America: Origins and Cultural Politics. Johns Hopkins University Press, Baltimore

Stone C (1974) Should Trees Have Standings? Toward Legal Rights for Natural Objects. Tioga Publishing Co, Palo Alto

Swyngedouw E (2007) Impossible 'sustainability' and the postpolitical condition. In: Krueger R, Gibbs D (eds) The Sustainable Development Paradox. Guilford Press, New York

Tallacchini M C (1996) Diritto per la natura. Giappichelli Editore, Torino

Tallacchini M C (2009) From commitment to committees. In: Seminar: Knowledge in Question. A Symposium on Interrogating Knowledge and Questioning Science, 597, May (also see www.india-seminar.com)

U.N. World Commission on Environment and Development (1987) Our Common Future. Oxford University Press, Oxford

Vanderbilt T (2008) Traffic. Why We Drive the Way We Do. Knopf, New York

Zimmerman A (2010) Moral Epistemology. Routledge, New York

Žižek S (1999) Carl Schmitt in the age of post-politics. In: Mouffe C (ed) The Challenge of Carl Schmitt. Verso, London, pp 18–37

Žižek S (2002) Looking Awry: An Introduction to Jacques Lacan through Popular Culture. MIT Press, Cambridge. (Originally published 1992)

Žižek S (2006) The Universal Exception. Continuum Books, London

Index

MIX
Papier aus verantwortungsvollen Quellen
Paper from responsible sources
FSC® C105338

Printed by Books on Demand, Germany

IDS Bibliotheken Bern
BM 1 788 462